# Information Practice
# in Science and Technology:
# Evolving Challenges
# and New Directions

*Information Practice in Science and Technology: Evolving Challenges and New Directions* has been co-published simultaneously as *Science & Technology Libraries*, Volume 21, Numbers 1/2 2001.

# Information Practice
# in Science and Technology:
# Evolving Challenges
# and New Directions

Mary C. Schlembach
Editor

*Information Practice in Science and Technology: Evolving Challenges and New Directions* has been co-published simultaneously as *Science & Technology Libraries*, Volume 21, Numbers 1/2 2001.

Routledge
Taylor & Francis Group
NEW YORK AND LONDON

First published 2001 by The Haworth Information Press

*Information Practice in Science and Technology: Evolving Challenges and New Directions* has been co-published simultaneously as *Science & Technology Libraries*™, Volume 21, Numbers 1/2 2001.

Published 2018 by Routledge
711 Third Avenue, New York, NY 10017, USA
2 Park Square, Milton Park, Abingdon, Oxon OX14 4RN

*Routledge is an imprint of the Taylor & Francis Group, an informa business*

Cover design by Lora Wiggins.

**Library of Congress Cataloging-in-Publication Data**

Information practice in science and technology : evolving challenges and new directions / Mary C. Schlembach, editor.
    p. cm.
    Co-published simultaneously as Science & technology libraries, v. 21, nos. 1/2.
    ISBN 0-7890-2183-8 (alk. paper) – ISBN 0-7890-2184-6 (pbk. : alk. paper)
    1. Technical libraries. 2. Scientific libraries. 3. Libraries–Special collections–Electronic information resources. 4. Electronic information resources–Use studies. 5. Libraries and electronic publishing. I. Schlembach, Mary C. II. Science & technology libraries.

Z675.T3I46 2003

                       2003007639

ISBN 13: 978-0-7890-2184-7 (pbk)

# Information Practice in Science and Technology: Evolving Challenges and New Directions

## CONTENTS

Introduction    1
*Mary C. Schlembach*

Merging Science/Technology Libraries: A Valuable
Planning Option    3
*Barton Lessin*

Electronic Collections–Evolution and Strategies: Past, Present,
and Future    17
*Carol A. Brach*

Providing a Digital Portal to a Print Collection: A Case Study
for an Engineering Documents Collection    29
*Winnie S. Chan*
*Deborah Rhue*

Challenges for Engineering Libraries: Supporting Research
and Teaching in a Cross-Disciplinary Environment    43
*Linda G. Ackerson*

Information-Seeking Behavior of Academic Meteorologists
and the Role of Information Specialists    53
*Julie Hallmark*

Geology Librarianship: Current Trends and Challenges    65
*Lura E. Joseph*

Bridge Beyond the Walls: Two Outreach Models
at the University of California, Santa Cruz    87
*Catherine Soehner*
*Wei Wei*

Current Implementation of the DOI in STM Publishing    97
    *Cynthia L. Mader*

When Vendor Statistics Are Not Enough: Determining Use
    of Electronic Databases    119
    *Amy S. Van Epps*

Capturing Patron Selections from an Engineering Library
    Public Terminal Menu: An Analysis of Results    127
    *William H. Mischo*
    *Mary C. Schlembach*

Looking for Numbers with Meaning: Using Server Logs
    to Generate Web Site Usage Statistics at the University
    of Illinois Chemistry Library    139
    *Beth L. Tarr*

Challenges and Changes: A Review of Issues Surrounding
    the Digital Migration of Government Information    153
    *Robert Slater*

Index    163

# ABOUT THE EDITOR

**Mary C. Schlembach, MLS, BS** is Assistant Engineering Librarian for Digital Projects at the University of Illinois at Urbana-Champaign. She received a Certificate of Advanced Study in Library Automation from the School of Library and Informtion Science, University of Pittsburgh. Ms. Schlembach is the editor of the American Society for Engineering Education's Engineering Libraries Division publication *Union List of Technical Reports, Standards, and Patents in Engineering Libraries,* Fourth Edition, as well as *Electronic Resources and Services in Sci-Tech Libraries* (The Haworth Press, Inc.). She currently serves as Chair of ASEE's Publications Policy Committee.

# Introduction

*Imagination is more important than knowledge. Knowledge is lim-
ited. Imagination encircles the world.*

–Albert Einstein

There are a myriad of challenges facing science and technology libraries as we move into the 21st century. Among the critical challenges
facing Sci-Tech libraries (and actually all libraries) are: the need to perform detailed collection assessment and evaluation, particularly in regard to e-resource collections; the need to examine and provide
appropriate public services; and the need to develop strategies for the
adoption of new information technologies. This volume presents twelve
papers that address these key issues and attempt to provide both perspective and insight into these problems.

The volume begins with papers relating to collection assessment and
evaluation. Barton Lessin's article on "Merging Science/Technology
Libraries" provides an example of continuing to provide service to various patrons in these difficult economic times for libraries. Like many
other science and technology libraries, the University of Notre Dame's
Engineering Library has concentrated much of their new resources on
electronic collections. Carol Brach's article, "Electronic Collections–
Evolution and Strategies: Past, Present, and Future," looks at important
issues connected with developing electronic collections. Finally, imple-

[Haworth co-indexing entry note]: "Introduction." Schlembach, Mary C. Co-published simultaneously in
*Science & Technology Libraries* (The Haworth Information Press, an imprint of The Haworth Press, Inc.) Vol.
21, No. 1/2, 2001, pp. 1-2; and: *Information Practice in Science and Technology: Evolving Challenges and New
Directions* (ed: Mary C. Schlembach) The Haworth Information Press, an imprint of The Haworth Press, Inc.,
2001, pp. 1-2. Single or multiple copies of this article are available for a fee from The Haworth Document Delivery Service [1-800-HAWORTH, 9:00 a.m. - 5:00 p.m. (EST). E-mail address: docdelivery@haworthpress.
com].

http://www.haworthpress.com/store/product.asp?sku=J122
© 2001 by The Haworth Press, Inc. All rights reserved.
10.1300/J122v21n01_01

menting changes in the University of Illinois' Engineering Documents Collection has provided Winnie Chan and Deborah Rhue with opportunities and challenges which other libraries may experience with in-house document collections.

Examining and providing appropriate public services is important in all libraries. Linda Ackerson's article examines support offered by libraries to academic faculty research, especially in new cross-disciplinary subject areas. Following are two articles on the specific subject area information needs of meteorologists by Julie Hallmark, and geologists by Lura Joseph. Catherine Soehner and Wei Wei describe outreach methods employed at the University of California, Santa Cruz to connect with library patrons and demonstrate services offered by the library.

Developing strategies and adopting new information technologies is imperative for libraries faced with the current information explosion. Cynthia Mader's article on Digital Object Identifiers (DOIs) thoroughly describes the new technologies associated with archiving and linking electronic information. Three papers are included on how libraries are implementing and gathering usage statistics to best serve their patrons' needs. Amy Van Epps discusses usage of electronic databases, William Mischo and Mary Schlembach are collecting usage statistics from library public terminals, and Beth Tarr's article focuses on the transaction log usage statistics of electronic reserves. Finally, Robert Slater discusses some of the issues surrounding the proposals to provide all government documents through an electronic distribution system.

This volume has been designed to provide information that will help address some of the new challenges and changes facing science and technology libraries, but these topics are important to all libraries. It is our hope that our imaginations will help solve the challenges and create new changes.

*Mary C. Schlembach*

# Merging Science/Technology Libraries: A Valuable Planning Option

## Barton Lessin

**SUMMARY.** The closing of an academic science/technology library is generally thought of as an extreme to be avoided. While sustaining separate departmental libraries may be preferred there is good reason for academic library planners to consider the closing of one or more libraries in order to merge them with another or other libraries. Given changes in library technology and the publishing industry, a merger offers the potential of improved patron services while helping the library prepare for a greater emphasis on digital information. This paper explores issues concerning the future of academic science/technology libraries and why the merging of academic departmental libraries should be used as an option in planning. *[Article copies available for a fee from The Haworth Document Delivery Service: 1-800-HAWORTH. E-mail address: <docdelivery@haworthpress. com> Website: <http://www.HaworthPress.com> © 2001 by The Haworth Press, Inc. All rights reserved.]*

**KEYWORDS.** Library space, scholarly publication, centralization, unsustainability, departmental library mergers

Barton Lessin, MS, is Assistant Dean of Libraries, Wayne State University, Detroit, MI. He is a member of ALA and ASEE-ELD and served as chair of the ACRL Science and Technology Section in 2002-2003.

The author wishes to thank Vivian Palmer and Charlene Sizemore (Wayne State University Science and Engineering Library) for their help in preparing this article.

[Haworth co-indexing entry note]: "Merging Science/Technology Libraries: A Valuable Planning Option." Lessin, Barton. Co-published simultaneously in *Science & Technology Libraries* (The Haworth Information Press, an imprint of The Haworth Press, Inc.) Vol. 21, No. 1/2, 2001, pp. 3-15; and: *Information Practice in Science and Technology: Evolving Challenges and New Directions* (ed: Mary C. Schlembach) The Haworth Information Press, an imprint of The Haworth Press, Inc., 2001, pp. 3-15. Single or multiple copies of this article are available for a fee from The Haworth Document Delivery Service [1-800-HAWORTH, 9:00 a.m. - 5:00 p.m. (EST). E-mail address: docdelivery@haworthpress.com].

10.1300/J122v21n01_02

## *INTRODUCTION*

Science/technology librarians may feel trapped between the proverbial rock and a hard place. We face the "rock" of ever-increasing patron expectations for services and access while dealing with the "hard place" requirement to respond to a constantly growing body of literature in a rapidly changing publishing environment. These challenges can test the strength of even the most stouthearted among us. Miriam Drake in her presentation to the Elsevier Digital Seminar posed the right questions for our consideration. She asked, "What is the future of scholarly publication?" "Will scholarly journals, as we know them, survive, and for how long?" "How will scientific and technical information be distributed?" (Drake 2001). These are obviously important questions and typical of the discussions that are omnipresent within the science-technology-medical (STM) library community as we attempt to plan for the future. Virtually all aspects of the academic science/technology library will be affected by changes in technology and publishing including the options available for the housing of library collections and services. It is in this context that library space planners should make a special effort to consider how these changes make the merging of some departmental libraries a practical, effective, and useful approach for enhancing patron services and materials housing.

As if these issues were not sufficiently challenging, there is the on-going discussion concerning the apparent decline in the on-site use of libraries that has recently become a part of the academic landscape. Although science/technology libraries were not specifically mentioned in Scott Carlson's *Chronicle of Higher Education* article focusing on the issue of decreasing on-site use of academic libraries (Carlson 2001), it is hard to dismiss or easily assess the impact of his work on administrators, faculty, and funding agencies in regard to those libraries. It is not difficult to understand the potential attractiveness for an approach that suggests that ever-greater amounts of information are available on the Web at the same time that fewer people are using academic libraries and questions the validity of expenditures to expand existing libraries or create new ones. If the experience at Wayne State University is a guide, one should not discount the decline in on-site use of academic libraries. Each of the libraries at Wayne State University save the Adamany Undergraduate Library has experienced a downturn in on-site use during the past four years. The Science and Engineering Library experienced a drop of approximately 55% during that time. There may well be conflicting issues, behaviors, and perceptions at work here. Academic patrons continue to value the library as a campus "place" according to the results of the 2002 Libqual+™ survey (Association of Research Libraries 2002). It is also acknowledged that academic patrons are making significant use of electronic resources that allow them to pursue their research without entering the library. Over 60% of those responding to Libqual+ indicated that they used electronic resources either daily or weekly. The change in use patterns appears to have impacted on-site use without diminishing how patrons view the library, its importance to their work, or its value as a campus location.

How will changes in scholarly publication, purchasing, patron use patterns, and technology impact library structures of the future? What impact will these changes have on science/technology departmental or subject-specific libraries? What options should librarians consider when asked by administrators how we intend to manage when a library has finite and quickly filling or filled shelving space at the same time as we continue to make purchases to expand the existing collections?

## *UNSUSTAINABILITY*

Brian Hawkins presents the strong argument that the academic library as we know it cannot be sustained. His position is based on the impact of the extant scholarly publishing model, the vast amounts of information being produced, and the impossibility of institutions of higher education and their libraries to keep pace with this productivity and its associated costs. His discussion of library space costs is particularly relevant here as he argues that "costs associated with storing and archiving the information will bankrupt our institutions of higher education" (Hawkins 1998). As if to help demonstrate the global aspects of this challenge, librarians who participated in an early 1998 Organization for Economic Co-operation and Development (OECD) meeting cited virtually the same issues as a part of their report concerning the future design and organization of academic libraries (OECD 1998).

If one largely accepts Hawkins' position that our collective situation is desperate and requires substantive action, several issues become apparent. The first of these is that no institution, including those that are the best funded, has the wherewithal to change the status quo on its own. Only collaborative action among libraries, institutions of higher education, and with faculty can result in the breadth and depth of change that is needed. The other is that we must consider alternatives to the way academic libraries currently store and provide access to information resources. Kaser reminds us that historically there has been a challenge to identify funding for academic libraries in the United States. He noted that "Institutional histories are replete with chronic, recurring threnodies upon their inability to house their libraries satisfactorily because of the unavailability of adequate funding" (Kaser 1997). Hawkins might suggest that the reasons for these lamentations will become magnified as costs and information volume increase. This implies that the future of the academic science/technology library may feature a limited set of information resources at the point of patron contact, supported by the bulk of collections housed in remote storage facilities, and complemented by international digital archives made locally available. Such a change would directly impact how some researchers do scholarship as well as how librarians provide access to information and could well impact the physical nature of the academic science/technology library.

## CENTRALIZATION OR DECENTRALIZATION

Merger implies a centralization of resources that is achieved in full or in part when multiple science/technology libraries are brought together at a common location. The merging of two or more libraries helps to centralize the involved services, resources, and collections for the patrons they serve. At the fullest extent of this process, one might find the science/technology collection as a distinct or integrated part of a comprehensive library. This definition of merger is intentionally less broad in scope than that used by Sacks in her discussion of merging libraries at related albeit geographically separated institutions (Sacks 1994).

The debate concerning the use of a central library facility or centralized science library structure as opposed to departmental libraries is not new to academic librarianship. Suozzi and Kerbel cite nine characteristics of departmental libraries that help to identify this type of library service. These include an identifiable clientele, focused goals and objectives, a holistic view of service, close proximity to the primary user community, inter-related functions, collegiality and flexibility among staff, entrepreneurial management style, the ability to develop and personalize service, and identification by staff and patrons alike as a part of the academic unit served by the library (Suozzi & Kerbel 1992). Changes in technological capability and scholarly publishing now integrated with academic library processes influence these characteristics in ways that academic librarians did not anticipate ten years ago. The characteristic of "close physical proximity to the primary user community" is a particularly useful example in as much as proximity to end-users was undoubtedly a primary catalyst in the establishment of departmental libraries in the first place. How much closer to users can academic libraries place information than on the very desktops of users and at those places where individuals prefer to work? There is, admittedly, a long way to go before a significantly large percentage of STM information is digitized; some resources may remain available only in paper while some disciples will enjoy more rapid digitization processes than others. Nonetheless, it seems clear that for science/technology libraries the trend that has been seen toward digitization will only continue and accelerate. Proximity to the user community should likely be reconsidered in terms of how quickly and effectively libraries can move needed information including library services to the desktop. Proximity can also be achieved in other ways. Librarians who work in a centralized environment can arrange to be available to the faculty and students in a college or department on a regular part-time basis if that is found to benefit users. Some libraries have established programs to specifically accomplish this goal. The Virginia Polytechnic Institute and State University model represents one approach to deal with this very situation (Eustis, Maddux & Dana 1995). It allows the university to gain the advantages of a central library with centralized staffing, hours, collections, and services while directly and positively dealing with a lack of departmental libraries.

One meaningful way to help assess the viability of a merger may be to apply the Suozzi and Kerbel departmental library characteristics to a potential new post-merger environment. One might argue that each of the characteristics used to distinguish the departmental library is eminently transferable to a centralized environment if in doing so the change leaders make these characteristics organizational and functional priorities. For example, it is possible to organize in a manner to assure that a librarian with academic credentials in a specific subject supports the academic department(s) offering related degrees. This approach is practical and appropriate for either the departmental or the centralized science/technology library. It is possible to envision a centralized organization that includes each of the listed characteristics depending on the outlook and approach to service by the staff of that centralized library. In this manner, it is plausible to apply the strengths of the departmental library to the centralized library environment. This would imply that it is also possible to merge two or more libraries to form a facility with combined services and collections that has the potential of mirroring the characteristics of a departmental library while providing the benefits of a merged environment.

## CLOSING/MERGING THE LIBRARY

As described by Davidson (1992) and Dekeyser (1998) there are a number of reasons why a science/technology library might be closed. The need for the space occupied by the library for other university or college purposes certainly must be considered. As science/technology departmental libraries are generally located in close proximity to the institution's science and engineering offices, labs, and classrooms and with space on most campuses a valued commodity, librarians should not be surprised to hear rumors that a department is interested in "reclaiming" space currently occupied by the library. Those interested in such a change might argue, "Isn't all the information available on the Web?" and "Aren't patron use figures for the library down as more researchers do their literature work at their desktops?" Many faculty who have access to departmental libraries use and appreciate them. However, as more information is made available online the need for proximity to a given collection may be diminished for many individuals and disciplines involved with STM literature. This reduced need for access to print resources may well accelerate as user information gathering behaviors continue to evolve. As this happens discussions about alternate uses of library space may also gather speed. The Physics Department at Wayne State University serves as an example of this thinking. With the availability of online resources from the American Institute of Physics and the American Physical Society including the latter's PROLA (*Physical Review Online Archive*) product, the department's liaison to the library advised that the faculty of that department simply wanted and needed electronic access to the journal literature. Once this access was assured the primary reason for faculty visits to the library was minimized. No departmental library is involved in this case, although had there been a call for reconsideration of that space it would not have come as a surprise given the use pattern shown by the members of this department.

Other reasons for the closing of a science/technology departmental library include lack of staffing or financial support to insure adequate hours for access or professional assistance. Changes in the organization of the institution with resulting impact on the library might also be a catalyst for a closure. Expansion space may be a driving force behind a possible closure as well. It may prove impractical to expand a departmental library owing to lack of proximate space to do so. Conversely, expansion space may be available but not offer the opportunity for growth that is estimated to meet the needs of the library for an extended period. Building condition is yet another reason that a closure might be considered. If the condition of the library building is such that the cost of repair is particularly high, it may prove more viable to close an existing departmental library and merge the collection and services into another library elsewhere on the campus.

The benefits of a merger can also serve as instigators for the consideration of a closure. In addition to the benefits associated with staffing, hours, combined information resources, and services, centralized operations may offer the potential of facilitated technological enhancement, improved visibility for library services making them more obvious to a greater portion of the institution's community, and greater efficiency through the use of centralized technical service operations (Dekeyser, 1998).

Planners should keep in mind that mergers can take on any number of different approaches and that the merger itself does not imply anything about the eventual location of a library. One might include among planning choices the merger of two or more smaller libraries into one combined library that occupies:

- the larger of the two or more available library spaces
- the larger of the two or more spaces augmented with expansion space
- the larger of the two or more spaces augmented with off-site storage in an annex or another library
- an existing "new" space designated for the combined facility
- an existing space augmented with additional space to accommodate the merger
- a new space in a building under development
- a new space in a new library facility developed specifically for that combined library

A merger can also be used to move even a single library to:

- distinct space allocated in a comprehensive library
- full integration within a comprehensive library

Each of these options offers planners the opportunity to combine collections and services and holds the potential for the organization to take steps that result in better staffing coverage, upgraded seating, lighting and other features that improve the comfort level of library users as well as to allow them to benefit from technological enhancements.

There are many examples where academic science/technology departmental libraries have been closed and merged with other libraries. Included here are brief descriptions of three mergers that help to illustrate some of the listed reasons for closing departmental libraries and pursuing mergers.

One of the best-documented examples was the merger between the John Crerar Library and the University of Chicago (Cairns 1986; Swanson 1986, 1990). In this case, once an institutional merger had been concluded that resulted in the Crerar Library moving to the university, an architectural firm was retained to design a new library for the Crerar collections and science collections from the university. In the mid-1980s this merger offered the faculty improved access to the scientific literature with the placement of the new library on the science quadrangle. The biomedical staff benefited even further when the planning included underground passages to classrooms, hospitals, clinics, etc. Swanson wrote that it was easier to arrange the internal functions within the new library than would have been the case if centralized services provided from the Regenstein Library had to be replicated. The motivating factors for this closure/merger were the need to establish a new home for the Crerar collections at the University of Chicago and the university's interest in merging related science resources with the Crerar collections to provide enhanced service with the primary focus on university patrons. Kathleen Zar, current director of the "new" Crerar Library, recently explained that the University of Chicago initially limited the merger to include only the Crerar Library, the Billings Hospital departmental library, and parts of collections from several other departmental libraries at the university. She also offered that discussions are continuing that may lead to the merger of several other departmental libraries into the Crerar Library (Zar 2002).

At the University of Louisville, the Natural Science Library and the Engineering Library were merged in 1983 (Brinkman & Kersey 1986). The former library had outgrown its available space and this space was needed for another university purpose–a new computing center. One of the optional spaces identified as a possible new location for the Natural Science Library proved unavailable, as the University had determined that space would be used for the Art Department. A merger with the main library was considered, but the Natural Science faculty vetoed that option citing the dual reasons of space limitations in the Ekstrom Library and distance from the Natural Science and Chemistry buildings. This example is particularly pertinent since the space for the combined libraries was too small to allow for the on-site storage of all the holdings of both libraries. Patron needs had to be determined so that some resources could be placed into storage. In retrospect, this particular requirement for this merger might serve as a precursor to the downsizing mergers that may wait future academic library planning. A feature of this merger that illustrates the problem of limited hours at departmental libraries was the distribution of keys to faculty for access to the Natural Science Library. In the absence of staffing and hours sufficient to meet the patron need, keys were distributed to help bridge the gap. Once the newly merged library was operational it was no longer necessary to provide keys given a new schedule that offered increased access hours.

Librarians at the University of Cincinnati documented the merger that occurred there and resulted in the placement of the geology and physics collections into a new building (Wells & Spohn 1990). These authors noted that there were a number of reasons for the merger including scattered collections, physical facilities that had fallen into disrepair, and a lack of needed space. Here again, keys had been issued owing to limited hours of operation in the libraries; this practice was discontinued after the merger no doubt making some people (librarians and administrators) feel that the library was more secure and others (faculty) that they had lost a special privilege. The new facility had a combined staff that was more than twice the size of that of the individual libraries prior to the merger. Modulex® signage and Space Saver® compact shelving helped to make the new facility more functional and efficient than the pre-merger libraries.

An informal solicitation of science and technology librarians via a listserv list produced a group of current mergers that are underway or recently completed at academic libraries across the country. Although not yet documented in the literature in the same manner as those cited here these most recent examples suggest that for at least some academic library organizations, the merger of libraries with the intention of service improvement and other benefits is very much a part of the current planning scene. Jim Oliver, Chemistry Librarian, and Carole Armstrong, Assistant Director for Collections and Human Resources described one merger at Michigan State University. The resulting merged library holds between 80,000 and 100,000 volumes all in compact shelving following the closure of the Chemistry Department reading room. The new facility is physically connected to several science buildings and is viewed as a service enhancement by both the libraries and the impacted academic departments. Convenience was seen as a key factor in this change. The MSU merger was one where two libraries were merged with some of the collection being located in another facility. In this case, that is the main MSU library that holds some of the older bound periodical literature. The Chemistry Department recouped the previous library space for more labs. Oliver explained that while there was not a financial savings from a staff reduction, the merger pulled existing staff together at the centralized location. Those departments served by the merged library now have the benefit of two full-time librarians rather than the one that was a part of each of the earlier distributed libraries. Oliver suggested that some saving is expected from the merger with the elimination of duplicate serial titles.

## *DECISION TO MERGE*

The library space planner has a number of options in determining steps to meet the space needs of a library. An existing science/technology library may be a candidate for enhancement to meet its space need. This change process may include renovation, remodeling, or even restoration depending upon the nature of the library and its importance to the institution (Bucher & Madrid 1996). These approaches are not mutually exclusive. In some cases, an addition is the approach that is required to meet the established goal and, of course, there is the option of building a new structure. A merger of libraries may combine a move of resources into an existing library, or an existing library with an addition, or into an existing library that needs remodeling or renovation, or as in the example from the University of Cincinnati into a new facility built specifically to meet the needs of the merged libraries. A merger in and of itself need not imply new construction, although this can be and is frequently one outcome of the planning process.

The choice to merge libraries rather than retain a departmental library and expand or otherwise change it will be determined by a number of factors. These factors supplement those referred to earlier having to do with library services and storage capacity. At the heart of the decision may be the set of goals that the library establishes for its planning project. Peña and Parshall offer a process that may prove useful in establishing goals for the project and later explaining them to others (Peña & Parshall 2001). Matters of structure, interior flexibility, cost benefit, adaptability for expansion, adaptability for new technologies and so on must also be considered. This part of the decision making process has been greatly facilitated by a list of questions provided by Philip Leighton and David Weber (Leighton & Weber 1999). Although written specifically for the consideration of building additions, the fifteen questions posed in this list are relevant to library mergers as well. One needs only to ask the questions in the context of including a new collection or multiple collections in a space originally designated for a single collection to have them apply. Examples from the list include these questions. "Is the location of the existing structure a good one?" Whether one is considering an expansion or a merger this question is paramount. Even with its substantial impact, digital information does not eliminate the need to assure that physical libraries are well positioned and easily accessible. "Is the existing structure sound?" If there are structural problems with an existing facility holding a single library, how could that facility possibly be improved by the inclusion of additional collections, services, patrons, operations, and other features of the merged library? "When the present building was planned, was an addition contemplated?" This question is pertinent to the planning of a merger in that the combined collections could at some future date require additional space. Would the chosen location be able to accommodate an expansion or is the library space limited in a manner that would preclude this? Expansion as an option from the outset may eliminate a painful reality years after the merger. The space planner considering a merger as a possible solution to an extant service or storage problem would be well advised to carefully consider each item on the Leighton-Weber list before proceeding with additional steps in the planning process.

## CONCLUSION

The merging of academic science/technology libraries is an option that should be considered by library planners who are concerned with the future of libraries and the ability of academic institutions to sustain multiple departmental or discipline-focused libraries. Merging offers the potential of better use of existing staff and more hours of operational service than might be found in separate departmental libraries covering the same subject areas. The examples of completed mergers suggest that this is an important option that can increase efficiency and enhance library services without creating barriers to access.

For some and perhaps many librarians and faculty, departmental libraries continue to have a role within the campus environment. While the reasons for creating departmental libraries still apply, assessment will be required on a case-by-case basis as facility planners determine the importance and the ability of the institution to sustain departmental libraries into the future. The impact of the digital revolution will challenge the role of the departmental library by putting more information on the desktop during a period in which libraries, regardless of size, struggle to manage growing collections. As more space is needed and projects are reviewed, planners will need to consider all options available to assure the best possible service to patrons. Mergers may offer an opportunity for service enhancement through improved hours and staffing, shared support services, centralization of resources, and improved efficiency. These benefits are important tools for planning the libraries that we expect to remain viable and functional for the next twenty-five to fifty years and more. This planning needs to occur in order to meet the challenges of expanding resources, limited shelving space, the costs of building or renovating, and the impact of digital technology on patron needs.

The operative question in regard to the use of mergers may be this. Given what we know about the growing volume of STM information, the limitations of existing library space, the costs for expansion, and changes in scholarly communications, are departmental libraries sustainable? The answer to this question will vary significantly from institution to institution. It is likely to vary from one departmental library to another with some determined to be more important than others. It is clear that while the use of mergers for space, service, and storage benefits is not a new approach, the growth and accessibility of digital information makes the merger of science and technology libraries more feasible and potentially more valuable to the university community than at any time since the initiation of the departmental library as a concept.

## REFERENCES

Association of Research Libraries. 2002. *Libqual+™ Spring 2002 Aggregate Survey Results*. Washington, DC: Association of Research Libraries.

Brinkman, C. S., & Kersey, L. 1986. The merger of University of Louisville's engineering and natural science libraries: a study in good planning. *Kentucky Libraries* 50: 25-29.

Bucher, W., & Madrid, C. 1996. *Dictionary of building preservation*. New York: Preservation Press J. Wiley.

Cairns, P. M. 1986. Crerar/Chicago library merger. *Library Resources & Technical Services* 30: 126-136.

Carlson, S. 2001. The deserted library. *The Chronicle of Higher Education* November 16: A35-A38.

Dekeyser, R. 1998. *Arguments for Library Centralization in the Digital Era* (Meeting report). Paris, France: Programme on Educational Building, OECD Experts' Meeting on Libraries and Resource Centres for Tertiary Education.

Drake, M. A. 2001. Science, technology, and information. *The Journal of Academic Librarianship* 27(4): 260-262.

Eustis, J., Maddux, L., & Dana, S. M. 1995. Adapting information services to new realities: the collegiate librarian/information officer program at Virginia Tech placing librarians in individual colleges of Virginia Tech to provide customized on site services. *Virginia Librarian* 41: 13-16.

Hawkins, B. 1998. The Unsustainability of the Traditional Library and the Threat to Higher Education. In B. Hawkins & P. Battin (Eds.), *The Mirage of Continuity: Reconfiguring Academic Information Resources for the 21st Century*. Washington, DC: Council on Library and Information Resources.

Kaser, D. 1997. *The evolution of the American academic library building*. Lanham, MD: Scarecrow Press.

Leighton, P. D., & Weber, D. C. 1999. *Planning academic and research library buildings,* 3rd ed. Chicago: American Library Association.

OECD. 1998. *Experts' Meeting on Libraries and Resource Centres for Tertiary Education by the Programme on Educational Building and the Programme for Institutional Management in Higher Education. Final Report.* Paris: OECD.

Peña, W., & Parshall, S. 2001. *Problem seeking: an architectural programming primer,* 4th ed. New York: J. Wiley & Sons.

Sacks, P. 1994. Merging Library Collections and Learning-Resource Technologies. In J. J. Martin & J. E. Samuels (Eds.), *Merging Colleges for Mutual Growth*. Baltimore and London: Johns Hopkins University Press.

Suozzi, P. A., & Kerbel, S. S. 1992. The organizational misfits. *College & Research Libraries* 53: 513-522.

Swanson, P. 1986. The John Crerar Library of the University of Chicago. *Science & Technology Libraries* 7:31-43.

Swanson, P. 1990. The politics of library buildings: a case study of the John Crerar Library. *IATUL Quarterly* 4:115-121.

Wells, M. S., & Spohn, R. A. 1990. Planning, implementation, and benefits of merging the geology and physics libraries into a combined renovated facility at the University of Cincinnati. *Paper presented at the Proceedings of the Twenty-Fifth Meeting of the Geoscience Information Society.*

Zar, K. 2002. Personal communication with author.

# Electronic Collections–
# Evolution and Strategies:
# Past, Present, and Future

Carol A. Brach

**SUMMARY.** Patron access to electronic collections through the Notre Dame campus network via the Web is seamless. A look at the important issues libraries are dealing with, as well as a discussion of the challenges of operating in a digital environment, are presented. *[Article copies available for a fee from The Haworth Document Delivery Service: 1-800-HAWORTH. E-mail address: <docdelivery@haworthpress.com> Website: <http://www.HaworthPress. com> © 2001 by The Haworth Press, Inc. All rights reserved.]*

**KEYWORDS.** Electronic collection, licensing agreements, publisher agreements

The University of Notre Dame has a growing collection of electronic journals, over 4,000 as of December 2001. More than 1,100 of these are in the subject area of engineering. Slower growth is being experienced in electronic reference materials. Mining the Internet for new electronic reference resources that are not "traditional" library materials continues to be an important endeavor for librarians in our role as selectors and collectors of important information. This paper will discuss issues asso-

Carol A. Brach, MLS, is Engineering Librarian, University of Notre Dame, 149 Fitzpatrick Hall of Engineering, Notre Dame, IN 46556.

[Haworth co-indexing entry note]: "Electronic Collections–Evolution and Strategies: Past, Present, and Future." Brach, Carol A. Co-published simultaneously in *Science & Technology Libraries* (The Haworth Information Press, an imprint of The Haworth Press, Inc.) Vol. 21, No. 1/2, 2001, pp. 17-27; and: *Information Practice in Science and Technology: Evolving Challenges and New Directions* (ed: Mary C. Schlembach) The Haworth Information Press, an imprint of The Haworth Press, Inc., 2001, pp. 17-27. Single or multiple copies of this article are available for a fee from The Haworth Document Delivery Service [1-800-HAWORTH, 9:00 a.m. - 5:00 p.m. (EST). E-mail address: docdelivery@haworthpress.com].

10.1300/J122v21n01_03

ciated with the selection and delivery of electronic journals to the desktops of faculty and students. While finding known journals in the digital environment has been simplified, finding known articles is still cumbersome, and finding new articles on a subject has never been more complicated. What will be presented here is an insider's view of the successes libraries have achieved, the associated issues that libraries are dealing with, and the challenges these electronic resources bring to our environment.

## *OUR SUCCESSES*

At Notre Dame, we have been able to amass a large number of electronic journals through subscribing to publisher packages. The value and sustainability of this practice has become debatable due to the channeling of a very large part of our collection budget to these publishers. They include: Academic Press (now Elsevier), Elsevier, Institute of Electronic and Electrical Engineers (IEEE), Institute of Physics (IOP), American Physical Society (APS), JSTOR, Kluwer, SIAM, Springer, and Wiley Interscience. We are working very hard to provide online access to all the important literature that is available in digital format. However, some smaller publishers and societies are trying to catch up with the larger publishers and there are troubling questions that arise such as whether Elsevier titles, for example, are over represented in collections of ejournals and will consequently be used more frequently than others. Additionally, there is the sense that the larger publishers are, by virtue of making their online journals so widely available, influencing the citation system so that journal impact factors are rising for the widely used titles, and lagging for even the best journals that are still making their way to the Web (Guedon 2001).

Libraries have put forth a tremendous effort to get individual title information into our online catalogs and provide links to electronic journals from there. We have constructed "gateways" to our electronic collections and portals to online information. This does not stop our patrons from going to the Web first to look for articles, while the carefully selected and very costly resources lie waiting for them at library Websites (Helman 2001).

In a recent article entitled, "The Design of Complex Library Web Sites," John Matylonek, Engineering Librarian at the Oregon State University Library states, "The increasing expectations of users for interfaces to lead directly, without undue hunting, to the information or service they need," is one issue that leads us to focus on the functionality of library Web sites. We need a way to handle "complex library research task flows due to the cross functionality and links between the online catalog, journal aggregator databases, electronic resources . . . It is no longer apparent how to do research with a tool (the library Website as a whole) that provides so many pathways for getting information." Matylonek goes on to propose analyses of critical paths and services, and gives ideas on how to eliminate obstacles that patrons can encounter when using library Websites (Matylonek 2001).

Patrons can now very easily go to a known journal title and download or print a full-text article from workstations located in the library, lab, office, home, or dorm room. Finding known articles is a little trickier because of the two step process of finding the journal first, then the article within the journal. Students frequently search our online catalog for specific journal articles, typing in the article title or author, and need to be educated to use the correct approach in order to be successful. This error is an instructional issue. In addition to retrieving known articles, one can now use electronic abstracting and indexing (A&I) services to search for new (unknown) articles, then retrieve them through links to the full-text of the article which are set up within the A&I service. The linking operation takes behind-the-scenes library intervention to set up, and has been very well received by users. One can do a search in *Web of Science,* for example, and link to an article by clicking on the "full-text" button that appears on the screen of each extended citation for which the library has established an online link. Patrons can also use search engines within collections of ejournals. For example, one can go directly to IEEE Xplore and do a search for items on particular subjects or by particular authors at the site, retrieving only IEEE articles. It is clear there are lots of options these days and can be a source of confusion for patrons.

## THE PROCESS

Libraries are buying publisher packages, linking individual "free with print" titles, and in general trying to find the best way to make everything available as quickly and efficiently as possible. The following is an outline of the steps that typically occur in the process:

- Identification of journal or package
- Negotiation (individually or through consortia) of access cost and other important restrictions and details

- Licensing contract must be put into place
- Journal or package of titles are added into the library collection and online catalog
- Delivery of journal(s) to the patron through the Web or local loading

While it is true that reliability and robustness of publisher and society or association sites are more a problem of the past, it is still a struggle to upgrade and maintain library computer equipment (RAM, etc.), and to provide for printing (including color printing when necessary) or saving of files. A more startling reality is the fact that "Libraries no longer own anything," and find themselves in the position of identifying and providing access only to "legitimate users" (Guedon 2001). Although most libraries try to negotiate usage of electronic resources to include those who may "walk in" to the library, the long held concept of free public usage of library materials is becoming harder and harder to provide. As a result of these new restrictions, information "haves" and "have-nots" are evolving. Third world countries and many smaller institutions in countries such as ours are the victims as access becomes increasingly dependent on finances, technology, the availability of equipment, and the availability of library resources such as staffing and computing power.

Another important detail of the new electronic environment deals with quality control and whether the content in the online version of a journal matches exactly with the content of the print version. It varies from case to case whether, for example, the values on a graph are readable online, whether letters and editorials appear in the online version, and if information on the cover is included. Some electronic versions of journals that also appear in print even have different ISSNs and are considered to be completely different publications. The problems associated with the possible differences between print and electronic versions of journals do not even address the fact that ejournals have capabilities that surpass print capabilities; for example, multimedia files can be included in ejournals. The close ties that print journals exert on their digital counterparts limit the digital format's evolution.

Licensing is a whole other quagmire. Presently, no standards exist for licenses and each must be examined individually by libraries who hope to offer online access to their patrons for publisher packages and individual titles. Staff time spent on licensing issues continues to be high. New licenses have to be negotiated by direct contact with the vendor; often the language must be revised and copyright details worked out to make sure the license does not deny rights granted under current copyright law including issues such as fair use, educational, and library exemptions (Brennan 1997). In most cases, signatures are required of both parties at the end of the negotiation process. Some publishers, for example, will even promise ownership of content when fees for access are paid from year to year, but are nebulous about how that might be achieved should the need arise. Most often libraries seek site licenses for access so that we can deliver electronic resources to our users on campus or to our distance users through proxy servers. We still struggle with publishers who will only allow access from one workstation on campus. Most libraries have experienced difficulties with publisher negotiations. When they have the opportunity, libraries have enlisted the help of campus legal resources to help with these negotiations. This, of course, adds to the cost of electronic resources. The Association of Research Libraries has a Website (http://www.arl.org/scomm/licensing/) with many licensing resources which can be helpful. There one can find licensing principles and examples of what some of the larger institutions such as MIT and the University of Texas recommend. Yale also has a site entitled LIBLICENSE, http://www.library.yale.edu/~llicense/ index.shtml, which is a resource of great value for librarians.

What direction do licensing agreements seem to be going? It is hoped that librarians can influence the future direction of licensing in order to minimize staff time and resources spent on the process. The following elements could be compiled into a single resource and become standardized: (1) common templates/boilerplates, (2) common principles, (3) common vocabulary, (4) blanket licenses, and (5) third party brokering (Bosch 1998). As Marian Burright, Science Librarian, George Mason University, a member of the STS Subject and Bibliographic Access Committee states, "a single solution may not be adequate" (Burright 1999). Examples of actual text of licenses can be found at http://www. library.yale.edu/~llicense/publishers.shtml.

## TRADITIONAL LIBRARY CONCERNS

So, what has become of traditional library concerns such as access to important scholarly content which has been carefully selected jointly by librarians and faculty who are experts in their respective disciplines? What about resource sharing? What traditional library services are evolving into electronic services?

To address these concerns, one can look to the some of the new endeavors with which libraries are involved in regard to scholarly publishing. Access presents challenges on a number of fronts. Most importantly, we are concerned with providing public access to all legitimate scholarly publications and information. Initiatives such as BioOne and SPARC are valued and supported. BioOne is enabling smaller publishers to migrate their publishing operations over to electronic publishing, which requires an initial investment that is out of reach for some (Alexander and Goodyear 2000). In contrast to that, the Scholarly Publishing and Academic Resources Coalition (SPARC) is providing a venue for new electronic publications that are more cost efficient alternatives to high priced journals (Assoc. of Research Libraries 2001). Libraries are taking the lead with these and other digital initiatives in order to contribute to the efforts that are occurring on many fronts during this digital revolution.

On the technical frontier, libraries are working to standardize links from cited articles to their "digital full-text representations" through the Digital Object Identifier (DOI) (Mischo 2001). This is a unique identifier of digital content and also a system to provide access to the content. CrossRef, a project and a service, is a central database where DOIs are being deposited. The database contains no full-text content, only the links. To date, there are 101 publishers participating in CrossRef, accounting for over 5,742 journals with over 4 million article records in the database. There are also a number of affiliates, library affiliates and associated organizations participating in CrossRef (PILA 2001). However, the dream of a multipublisher, metadata depository is not yet on the horizon. As it stands, CrossRef metadata only includes the first author, for example, and has no controlled vocabulary and thus has limited retrieval capabilities.

Access in perpetuity is yet another concern. While some publishers will agree to "permanent" access to content when libraries pay for access over a period of years, the details about this are sketchy. How we might obtain and archive the data is undetermined. "Subscriptions" to online content are inherently different from the known scenario of obtaining new volumes each year, and binding and retaining them as part of a permanent collection.

Local archiving of electronic journal articles by libraries or consortia has been suggested as a reliable solution to this dilemma (Burright 1999). The debate continues as to who might take responsibility for archiving and, in general, publishers have made no promises to the library community that they will archive what they have published online. It is surmised that additional costs will be involved should the publishing community take the eventual lead in this issue since profitability continues to be most important to publishers. Present technology is not adequate to assure access in perpetuity to ejournals. Therefore, archiving remains an important unresolved issue.

## THE FUTURE

The Digital Library Federation has taken a leading role in the ongoing investigation of digital archiving by supporting joint projects between libraries and publishers. Yale has teamed up with Elsevier; Harvard with Blackwell, Wiley and the University of Chicago Press; and Cornell is taking a subject approach and is working with as many as a dozen publishers of core agricultural journals. Funding comes from members and grants. A summary of the projects and progress can be found at http://www.diglib.org/preserve/ejpreps.htm#cornell.

Libraries are definitely concerned with the preservation of digital content and are working in conjunction with publishers on possible long-term solutions, but a different perspective is emerging from the publishers' point of view. In a recent interview in *Information Today*, Derk Haank, Elsevier Science Chairman, gives some details on their idea of a model for ongoing electronic access to their 20% of the market by stating that they want to "transform a system where only a few libraries pay for a few journals to where all libraries pay something for access to everything." Elsevier plans to fund the whole scenario with the proceeds from library subscriptions, which means we are funding what they are driving and, ultimately, what they consider to be most important. What will happen to that effort when we can no longer afford to subscribe? Haank also states that "the real impact of electronic publishing is not that the costs come down. If anything, the costs go up dramatically" (Kaser 2002). Their thinking is based on a profit model which they may struggle to sustain. What will happen to electronic back files and will publishers decide to split those off from the current years and begin charging twice–once for current years and another charge for maintaining back files? What would prevent them from doing that? An assurance has been officially offered by Karen Hunter, Senior Vice President of Elsevier Science, in a posting to the Yale LIBLICENSE listserv on November 21, 1999, "Elsevier Science has recently announced our commitment to perpetual archiving and our assurance that we will not permit the database to be dismantled or otherwise taken down without depositing copies in library-approved facilities. We are already investigating some of the options for deposit internationally . . . One of the things we need to work out is how and when the hand-off of the archive takes place" (Hunter 1999).

Archiving of older issues of print journals remains an important consideration to libraries that may give up print subscriptions in order to pay for current online subscriptions and thus no longer even have the ability to bind and store individual issues and volumes. Libraries do not currently have group strategies to archive and access collections which are duplicated in many libraries across the nation. In addition, there are simply no guarantees from publishers that older volumes will be digitized or that older volumes of digitized journals will remain accessible as archives build up and grow larger and larger. As dusty as they may get, older volumes of journals are still important to researchers. The fact also remains that many publishers may not have the means or do not choose to evolve to digital for any but the most current issues and so back files may languish on library shelves, some destined never to become electronic files.

Not necessarily the last and certainly not the least of the concerns is a possible unexpected consequence for libraries that can be characterized as "core collecting." By this I mean that libraries that must deal with limited resources in a publishing world where not every journal is electronic and not every back file is available, will be making choices on what to subscribe and provide access to based on the tried and true model of providing users with "core" titles. As funds are stretched thinner, all but the core journals may be cut by most libraries, leaving national collections, formerly rich and robust, with 10,000 copies of the same primary journals and very few copies of peripheral yet important journals. More than ever, cooperation among libraries will grow in importance, with state and regional collection centers becoming necessary. Even with cooperatives and consortiums in place, primary journals will be required in nearly all libraries. Alternatively, careful collection planning among dispersed groups of libraries could evolve. The main point is that collections will be changing along with how information is accessed. New services will have to be put into place to answer information needs that have, historically, been handled locally. As the idea of "place" becomes less important than being able to obtain information that is needed, new and improved virtual services will have to evolve as well. User expectations continue to be important to libraries and we continue to seek to identify what those are.

The Center for Research Libraries (CRL) has a number of collection programs exploring access to and preservation of scholarly journals that may prove to be extremely valuable in the future. Their Website, located at http://wwwcrl.uchicago.edu/info/intjournl.htm, lists several initiatives currently being considered or in progress for scholarly journals including:

- Expanding the availability of rarely held scientific titles by employing a centralized cooperative collection development model to pool funds to acquire carefully selected titles to house at CRL under shared ownership (Scientific Research Materials Project).
- Making available new collections of serials in electronic format.
- Exploring the potential to assist member libraries faced with ongoing journals cancellation projects.
- Expanding the CRL retrospective serial titles archive by partnering with emerging organizations providing electronic access to retrospective issues of journals to maintain a backstopping security archive (CRL 1998).

A new strategy that is evolving at Notre Dame is just-in-time delivery of specific research materials our patrons may need; that is, delivery of materials whether or not we own them as simply and quickly as possible. This is not a new idea. However, this strategy has the implication that collections and collecting will be de-emphasized, especially for materials and for disciplines that are not considered to be Notre Dame subject strengths. As libraries seek strategies to make our dollars stretch just as far as we can, choices like this one may become necessary alternatives to the collection-based ways we have done things in the past. As demand for items that were cancelled or that we never owned goes up and collection dollars are transferred to document delivery budget lines, what will the consequences of such decisions be locally and nationally? How will our peer-based rankings, such as those in the North American Title Count, be affected? This remains to be seen. Further, who will our future document delivery vendors be? Will they be publishers or will the interlibrary loan function, that is organized and based in libraries, grow and change to meet the demands of short fulfillment cycles and flourish as demand grows?

## NEW AND OLD CHALLENGES

How does our changing environment affect what we do and how will we respond to the new digital initiatives? Library instruction, skills training, education, and information literacy are all being rethought so that users can have the greatest advantage when seeking information inside or outside the library. Publicity, not an area in which libraries are particularly strong, is now more important than ever. Libraries are amassing collections beyond our walls and are struggling to connect our users with the plethora of resources that are now available to them. All the while, librarians are rethinking "how" people are seeking information and how libraries might assist them by continually redesigning our Websites and other information resources.

An old focus has taken on new importance to librarians–information literacy. Often considered a higher level activity during times when focusing on the "basics" seemed expedient, the current focus in library instruction necessitates increased attention to information literacy. We feel that success, and indeed perhaps survival of patrons in our complex and information rich environment, depends on the development of higher-level retrieval skills. Indeed many students do not even realize that library resources exist beyond what is available via a search on the Internet. Many do not even know that they will not find our resources through "Google." Many need help in evaluating other resources they find on the Internet and many do not even know that we spend time selecting good quality sites for them to use. The two streams of Internet resources and library resources rarely intersect.

Ironically, focus group interviews recently conducted at Notre Dame have revealed that what our patrons want is more simplified access to all that we have to offer them, from items in our library catalog to the full-text of journal articles. They want to be able to search all of our resources with an interface as comprehensive and simple as "Google." While we are busy planning complex instruction strategies, our patrons want us to deliver what we have in the same way that the Internet delivers what it has. We want to give them what *we* think they need, while they are asking us for what *they* think they need. How can this dilemma be solved?

Electronic services are evolving as alternatives to formerly mediated services such as patron initiated interlibrary loans and electronic document delivery. More and more of our users are filling out Web forms to place requests for information resources we do not "own" and as we move towards more seamless access to resources, we now have the ability to send documents directly to patrons via their email accounts. Copying and faxing services are giving way to scanning and attaching files to email messages. All in all, even when service possibilities exist, libraries are struggling to reassign staff to new duties and to educate ourselves about new delivery methods all the while being mindful of the ever changing copyright implications of the new services. As a body, we have the collections. The challenges are to preserve and deliver our resources seamlessly and quickly to researchers in any location. As we all move forward together in this new digital world, librarians are committed to continue to organize "the literature," regardless of what form it may take, how we may provide access to it, and the ways in which we will preserve it.

## REFERENCES

Alexander, A., Goodyear, M. 2000. The Development of BioOne: Changing the Role of Research Libraries in Scholarly Communication. *JEP: Journal of Electronic Publishing* 5(3):March. Available at: http://www.press.umich.edu/jep/05-03/alexander.html.

Association of Research Libraries. 2002. *SPARC: The Scholarly Publishing and Academic Resources Coalition.* Available at: http://www.arl.org/sparc/core/index.asp?page=a0.

Bosch, S. 1998. Licensing information: where can we go from here? *Library Acquisitions: Practice & Theory* 22(1):45-47.

Brennan, P. et al. 1997. Licensing Electronic Resources: Strategic and Practical Considerations for Signing Electronic Information Delivery Agreements. Available at: http://www.arl.org/scomm/licensing/licbooklet.html.

Buckley, C., Burright, M., Prendergast, A., Sapon-White, R., and Taylor, A. 1999. Electronic Publishing of Scholarly Journals: A Bibliographic Essay of Current Issues by the STS Subject and Bibliographic Access Committee. *Issues in Science & Technology Librarianship* 22(Spring). Available at: http://www.istl.org/99-spring/article4.html.

Center for Research Libraries. 1998. Scholarly Journals. Available at: http://wwwcrl.uchicago.edu/info/intjournl.htm.

Guedon, J. C. 2001. Beyond Core Journals and Licenses: The Paths to Reform Scientific Publishing. *ARL: A Bimonthly Report on Research Library Issues and Actions from ARL, CNI, and SPARC* 218(October):1-8. Available at: http://www.arl.org/newsltr/218/guedon.html.

Helman, D., Horowitz, L. R. 2001. Focusing on the User for Improved Service Quality. *Science & Technology Libraries* 19(3-4):207-219.

Kaser, Dick. 2002. Ghost in a bottle; Elsevier Science Chairman Derk Haank responds to the Public Library of Science initiative. *Information Today* 19(2):1, 46.

Matylonek, John. 2001. The Design of Complex Library Web Sites. *Transforming Traditional Libraries* 1(2). Available at: http://www.lib.usf.edu/~mdibble/ttl/Clustr1.htm.

Mischo, W. H. 2001. The Digital Engineering Library: Current Technologies and Challenges. *Science & Technology Libraries* 19(3-4):129-145.

Publishers International Linking Association (PILA). 2000. *CrossRef* Available at: http://www.crossref.org/.

# Providing a Digital Portal
# to a Print Collection:
# A Case Study
# for an Engineering Documents Collection

Winnie S. Chan

Deborah Rhue

**SUMMARY.** Unpublished reports are a valuable source of information that is essential to both academic and industrial research. The Engineering Documents Center collection at the Grainger Engineering Library of the University of Illinois at Urbana-Champaign makes technical documents produced by the College of Engineering faculty available through a locally developed Web interface. This paper describes a study conducted to enhance the interface into a digital portal to the print collection. A comparison of the Web interface (a locally developed model) and CONTENTdm

Winnie S. Chan is Assistant Engineering Librarian and Assistant Professor of Library Administration, University of Illinois at Urbana-Champaign, Urbana, IL (E-mail: w-chan2@uiuc.edu). Deborah Rhue is a graduate student in the Graduate School of Library and Information Sciences, University of Illinois at Urbana-Champaign, Urbana, IL (E-mail: rhue@uiuc.edu).

The authors would like to thank Robert Slater for his assistance in preparing this manuscript.

[Haworth co-indexing entry note]: "Providing a Digital Portal to a Print Collection: A Case Study for an Engineering Documents Collection." Chan, Winnie S., and Deborah Rhue. Co-published simultaneously in *Science & Technology Libraries* (The Haworth Information Press, an imprint of The Haworth Press, Inc.) Vol. 21, No. 1/2, 2001, pp. 29-42; and: *Information Practice in Science and Technology: Evolving Challenges and New Directions* (ed: Mary C. Schlembach) The Haworth Information Press, an imprint of The Haworth Press, Inc., 2001, pp. 29-42. Single or multiple copies of this article are available for a fee from The Haworth Document Delivery Service [1-800-HAWORTH, 9:00 a.m. - 5:00 p.m. (EST). E-mail address: docdelivery@haworthpress.com].

10.1300/J122v21n01_04

is provided. The decision-making process and the technical considerations for creating suitable text images are also discussed. *[Article copies available for a fee from The Haworth Document Delivery Service: 1-800-HAWORTH. E-mail address: <docdelivery@haworthpress.com> Website: <http://www.Haworth Press.com> © 2001 by The Haworth Press, Inc. All rights reserved.]*

**KEYWORDS.** Unpublished reports, print collection, digital content, images

## INTRODUCTION

Information seekers in the fields of science and technology rely on the use of many bibliographic tools to locate and access scientific serial literature, conference proceedings and unpublished reports. Unpublished reports are useful and essential to both academic and industrial research because they convey new developments of scientific and technical research and provide a forum for peer information exchange. Also, unpublished reports often contain more details than can be found in published materials (Zulu 1993).

On the national scene, a network of clearinghouses such as the U.S. Government Printing Office (GPO), the National Technical Information Service (NTIS), and the Defense Technical Information Center (DTIC) and its Scientific and Technical Information Network (Public STINET), make available to business and industry the results of government-funded research and development. At the local level, unpublished reports are generally characterized as the type of *grey literature* that is not available through normal commercial publishers and has low print runs (Thompson 2001). Because of these inherent problems, access to unpublished reports is greatly hindered and depends heavily upon the infrastructure of the local institution within which document delivery services are performed.

## THE ENGINEERING DOCUMENTS COLLECTION AT THE UNIVERSITY OF ILLINOIS

At the University of Illinois at Urbana-Champaign (UIUC), the College of Engineering created a documents center in 1966 to serve as a repository for the research findings produced at the university's laboratories and reported by faculty members. The Engineering Documents Center, as it was known since its inception, played a major role in providing document delivery service for the College. The collection has been growing at a rate of 100-200 reports annually, totaling over 14,000 titles by 1999. Many of these titles are not available elsewhere but are nonetheless referenced to and requested by researchers, practicing engineers and scholars from around the world.

With the retirement of the Documents Librarian at the Center in 2000, the Grainger Engineering Library absorbed the responsibility of managing the collection, known henceforth as the Engineering Documents Collection (Chan 2002). The collection database, converted into a Microsoft Access database from a dBase format, forms a part of Grainger's locally developed Web-based database family for enhanced access to specialized library resources (Mischo and Schlembach 1999). As such, the customized Web applications for the Engineering Documents Collection include a Web search form, single/multiple search results display options and an online form for submitting document requests. Search options include keyword searching for author, title, subject, departmental series, research laboratories, contract numbers, sponsors, and document accession numbers.

Although the Engineering Documents Collection is housed at the Grainger Engineering Library, it is a separate collection from the one represented by the UIUC Library online catalog. The resource is free to the University community, but there is a nominal cost recovery charge for document delivery service imposed on document requests from non-UIUC members. The expense covers the copyright fee, photocopy and postage costs.

## DIGITAL PORTAL
## TO THE ENGINEERING DOCUMENTS COLLECTION

As the Engineering Documents Collection grows, the concern for preservation as well as storage becomes a challenging archival issue. While many of the earlier technical reports are out of print or close to out of stock copies, current technical reports are available in both electronic and print formats. Our current Web interface to the Collection provides basic keyword searches (http://shiva.grainger.uiuc.edu/engdoc/opent1.asp). However, there is no program in place for delivering full-text files for document requests. In mid-2002, plans were underway to investigate making a digital portal to the Engineering Documents Collection to enhance its current Web interface.

Decisions to digitize a large collection such as the Engineering Documents Collection should reflect the demand for document delivery service. We began with studying changes in the infrastructure, namely,

- Chartering a progressive plan that allows gradual creation of scanned objects without losing full access to the entire print collection,
- Designing a Web interface that serves as a digital portal for access and document delivery service, and
- Developing a clearly defined step-by-step scanning workflow to produce best textual images cost efficiently.

## *SOURCES OF DIGITAL IMAGES*

The approach to digitizing an existing collection should take into consideration the size of the collection, regardless of the resources available and the impact on access. Determining what is to be digitized and in what order is a major preliminary step in this undertaking. We decided to have two separate procedures for preparing digital images of documents: one for creating Web images of the title page, abstract and the table of contents of the document, and one for creating full-text documents. With this approach, we allow access via the digital portal to the print collection while maintaining the current level of access and slowly growing our selection of full-text holdings.

As a pilot study, we selected 23 documents and created the Web images of title page, abstract and the table of contents to be displayed via the Web interface, which serves as the digital portal to the entire Engineering Documents Collection. Providing full-text documents is to be undertaken through different methods. The first step towards achieving this long-term goal is to seek full cooperation of the report producing departments and research centers. They would then furnish the electronic version of their current documents in addition to the print copy. As for the older, print-only reports, digitizing documents on demand is the best way to achieve our goal. This 'on-the-fly' option is well suited for delivering documents which are being requested repeatedly.

# CHOICE OF WEB INTERFACE

## *From the Web Interface at the Grainger Library Home Page*

With Web access to the collection already in place through the Grainger Library Home Page, the Engineering Documents Collection Web interface was further enhanced in the display of search results. A thumbnail of images for title page and abstract/or table of contents is added as part of the single record display screen. Figure 1 shows the display of a search result on Prof. Bullard and his reports on heat exchangers.

From this list, Document 2001-4009 is selected to display the full description of the record. Figure 2 shows the single record display of Document 2001-4009 with title page and abstract in the thumbnail area.

## *From the Web Interface by CONTENTdm*

To gauge the effectiveness of our approach, we compared our system to CONTENTdm Multimedia Archival Software. The same set of images was loaded into CONTENTdm. Figure 3 shows a search query form provided by CONTENTdm using Authors as the browse term operator. The results for a search query on Prof. Bullard and his reports on heat exchangers produce a line of thumbnail images at the bottom half of the screen.

FIGURE 1. Search Results from the Grainger Web Interface

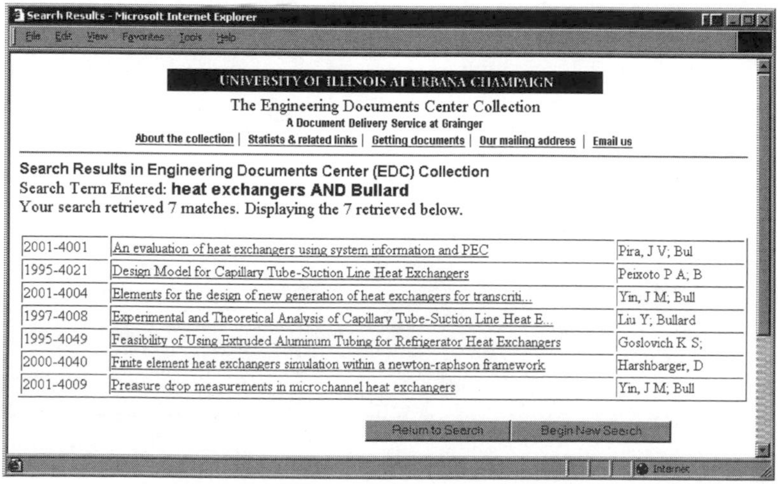

FIGURE 2. Single Record Display with Thumbnail Images

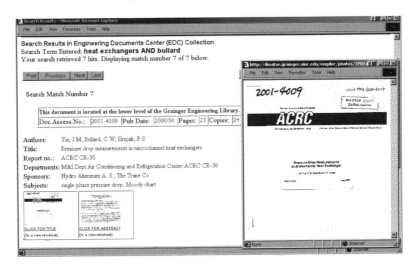

FIGURE 3. Search Results from CONTENTdm

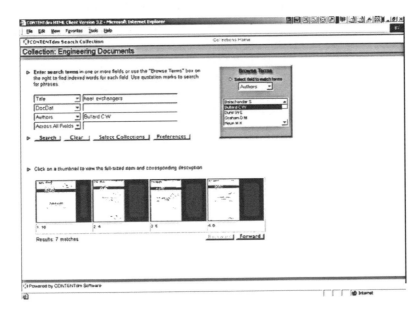

The subsequent image and description of the item displays are shown in Figure 4.

## Comparison of the Grainger Model versus CONTENTdm

Using a model employed across several locally developed databases, we enhanced the Web interface to display the scanned pages in a thumbnail table on the search result screen. Upon a click of the mouse, the image is opened in a new window with the description of the search record still in view. In the CONTENTdm model, the thumbnail table displays the search matches. The image and description are rendered separately, each with a click of the mouse. CONTENTdm obfuscates the digital contents more than the Grainger interface.

Most importantly, we compared the feature that has an impact on accessing a collection which contains a subset of digital contents. With the Grainger model, the search function remained intact, retrieving from the entire database matching records with or without embedded digital images. With CONTENTdm, each item in the collection must be associated with one or more images; therefore, accessibility to all records in a large collection such as the Engineering Documents Collection would not be possible until all digital images are available.

While CONTENTdm is the industry's premiere technology for digital media management, it is, however, incompatible with a print collection which contains mainly textual contents. The limitation of image-only

FIGURE 4. An Enlarged Thumbnail Image on the Left and a Record Description Image on the Right

components does not allow us to approach digitization feasibly while maintaining full access to the existing collection of over 15,000 reports.

## SCANNING WORKFLOW OVERVIEW

We experimented with the hardware and software available at hand to prepare the Web images of title page, abstract and the table of contents for our sample documents. Figure 5 illustrates the overview of the scanning workflow. The technical considerations of software in the scanning operation are detailed in the following sections.

## TECHNICAL CONSIDERATIONS OF SOFTWARE FOR SCANNED IMAGES

Initially, we decided to keep the original scan of our documents for archival purposes and provide a smaller version of the image for access by our patrons over the Internet. The need to keep the original scan was made evident by the amount of time it took to produce each scanned im-

FIGURE 5. Scanning Workflow

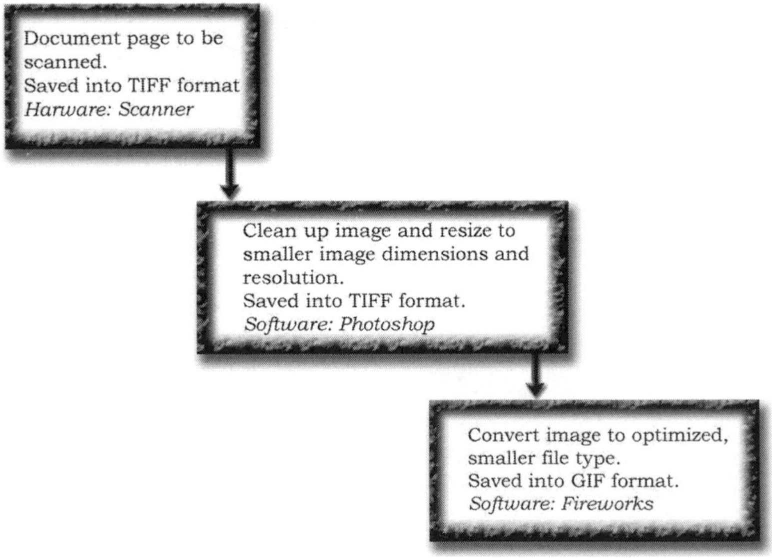

Document page to be scanned.
Saved into TIFF format
*Harware: Scanner*

Clean up image and resize to smaller image dimensions and resolution.
Saved into TIFF format.
*Software: Photoshop*

Convert image to optimized, smaller file type.
Saved into GIF format.
*Software: Fireworks*

age. Using an Epson flatbed scanner, we found that our capture image (ranging in size from 4″ × 8″ to 8″ × 10″; grayscale, 600 dpi resolution) took approximately 15 to 30 seconds for one scanning pass. Additional time was spent initiating the scanning software, doing the pre-scan, setting the scanning boundary box, and adding and removing the item from the scanner. The overall time required for a one-page scan approached 3 minutes. For each document selected for inclusion in this Web image project, we scanned the title page, the abstract, and/or the table of contents (usually about two pages).

With the TIFF scans in grayscale, high resolution and uncompressed format, we then moved on to determine how best to prepare these images for storage and delivery over the Internet. Our objective was to devise a configuration for production of a much lower resolution, highly compressed format. We chose 72 dots per inch (dpi) as our output resolution from Photoshop with a width of 650 pixels, which resulted in fairly readable on-screen displays with minimal scrolling (on computers with output resolutions set to 800*600 dpi or more) and a legible printed output. Then we experimented with the file compression versus image quality tradeoffs of the various browser viewable file formats, resulting in a choice between the JPG and GIF formats. We tried using Photoshop's bundled Web exporting program, ImageReady 3.0, and compared it to Macromedia's competing program, Fireworks 4.0. We found that Fireworks slightly outperformed ImageReady in converting our grayscale 72 dpi TIFF images into a web format. The JPEG format tended to be at least twice the file size for a comparably usable image in both programs. Even at the lowest setting, the JPEG format couldn't match the GIF file size. Since our TIFFs were grayscale and composed almost exclusively of clear text on a blank white background, the GIF format worked better. We took what we considered to be a representative sample of our documents and found that the only items in the documents besides text were occasional line drawings or simple illustrations. If we come across any photo-realistic images in the documents we are scanning, we may need to reconsider our output format to accommodate those exceptions.

Speckling and bleed-through of text from the backs of scanned pages occasionally distorted image quality and produced larger files. When we began to convert these images it was obvious that some measure of touch-up would be required to make the image text more legible in compressed versions, and to decrease the size of the files. We devised a two-phase process for this, which was eventually automated for convenient batch processing. The first phase was the clean-up phase, which was done in Adobe Photoshop 6.0. This extremely powerful software tool offers a host of image manipulation algorithms called filters, along with very impressive image re-sampling algorithms that allow large resolution/dimension images to be accurately scaled down. It is important to note that these steps are best used with *high resolution* (300 dpi or more, preferably 600 dpi) *grayscale* TIFF images. When working with color images, first use Photoshop's **Image → Mode → Grayscale** to convert the image to a grayscale mode. The steps we took in phase one (image clean-up) were as follows:

STEP 1  **Filter → Noise → Despeckle**

A precautionary procedure to remove unwanted speckling from older documents, although most of the documents in our collection are exceedingly clean. Although this step makes the image more presentable, it also reduces the eventual file size by removing the unnecessary information that would need to be in the image file to represent each and every speckle.

STEP 2  **Image → Adjust → Auto-Levels**

This step color balances the picture, which is more important for color images, but still possibly useful in a grayscale image.

STEP 3  **Image → Adjust → Auto-Contrast**

This will do just what it says, analyze the image and then make the blank paper a more continuous, bright white color, while taking the gray-to-black letters of the text and making them a more continuous dark color.

STEP 4  **Filter → Sharpen → Unsharp Mask → Amount: 150%, Radius 5 pixels, Threshold 10 levels**

This will make the most dramatic difference. It adjusts the bleed that accompanies the edge of any printed text and sharpens and hardens the text edges, making it more readable, but also reducing the overall amount of information that needs to be preserved in the image.

STEP 5 **Image → Image Size → [Document Size] Resolution 72 pixels/ inch, [Pixel Dimensions] width 650 pixels (Constrain Proportions checked; Resample Image: Bicubic checked)**

Please note that it is necessary to change the resolution first, as this affects the pixel width of the document.

At this point, we had a much smaller file that we saved as an uncompressed TIFF image. Resizing the image and lowering the resolution caused most of the change in file size. Once we had determined the exact procedures we wanted to use, and the setting for the **Unsharp Mask** filter, we used the **Actions** feature of Photoshop to create a macro (see Figure 6). Once our macro was created, we used Photoshop's batch processing feature to manipulate an entire folder of images, and then placed the cleaned-up, smaller TIFF images into a temporary folder.

With our smaller files in a temporary folder, we used Firework's batch processing feature to export all the images in our temporary folder into the final, finished image folder. For this processing, we used the

FIGURE 6. The Actions Palette in Photoshop

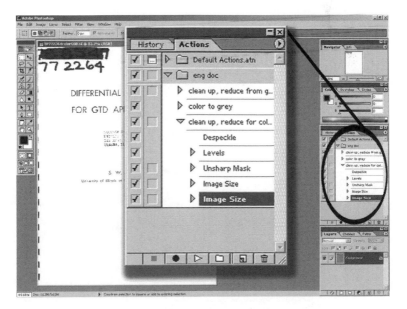

**GIF** file format with the following settings: **Uniform, 64 color palette** (which Fireworks automatically reduced to 4 to 8 colors), **Loss** 100%, **Dither** 0%, and **no transparency** (selecting either of the transparency modes, with the matte color set to white, only marginally reduced image size, so we decided to leave the background of the documents white, as opposed to transparent). The finished products were very legible files that were between 15 and 20 KB (see Figure 7).

Given this size, we could archive our entire collection in less than 600,000 KB, or 0.58 Gigabytes of hard drive space, which we found acceptable. We found that we could also rely on Fireworks Black & White export settings for GIFs to convert some color scans (mostly title pages that employed 3 or 4 color printing) to black and white GIFs, which had an incredibly small file size but often lost some of the text clarity in the conversion process.

## ARCHIVING CONSIDERATIONS

Our original scans resulted in uncompressed TIFF grayscale images with resolutions of 600 dpi. The resolution is the single most influential

FIGURE 7. Export Setting in Fireworks

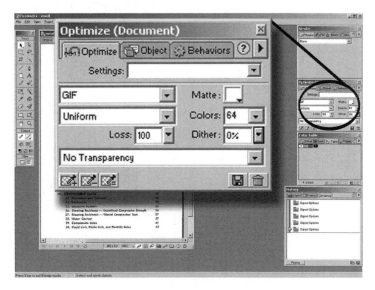

independent variable on the resultant file size of a scanned image. Our 8″ × 10″ 600 dpi grayscale scans resulted in files that approached, but rarely exceeded, 30 megabytes. Taking the 30 megabyte file size as the average (the worst case scenario), over 3000 images at this setting would fit on one 100 gigabyte hard drive disk, while a single-sided, single-layer DVD (4.7 gigabytes in size) would be able to archive more than 150 images per DVD.

The Engineering Documents Collection is comprised of over 14,000 documents, resulting in (for this Web image project) one scanned image of the title page and one scanned image of the abstract or the table of contents. This means we would need to archive 30,000 images, requiring at least 9 dedicated 100 gigabyte hard drives. As removable media, this would result in roughly 75 documents archived per DVD disk ([4.7Gb/DVD * 1024Mb/Gb]/[30Mb/image * (2 images/document)]), or approximately 200 DVDs. We decided that archiving at this level was an unwieldy task that would need to be re-evaluated from a cost-benefit perspective. Eventually, we decided to streamline our conversion efforts. We scanned the archival image directly into an OCR software package (e.g., ABBYYY FineReader), ran the OCR conversion, saved the unchecked OCR version, and then temporarily stored the archival TIFF image. As time allowed (and a reasonable number of archival images had been accumulated) the archival images were batch processed into much smaller, Web-portable image sizes/formats, which then became both the archival and presentational copy of the scanned image. The original high resolution TIFFs were deleted to conserve disk space.

## *CONCLUSION*

The value of unpublished reports as a source of useful information is essential to both basic and applied research in the fields of science and technology. Characterized as grey literature, they are often not easily accessible. At the University of Illinois at Urbana-Champaign, a unique collection of unpublished reports authored by the College of Engineering faulty has been made Web-accessible since 1999 from the Grainger Engineering Library. In order to further enhance the document delivery service, we embarked on a feasibility project to study two digital portals in place of the current Web interface. Taken into consideration are the size of the collection, the impact on accessibility, the infrastructure change, workflow, and the technical details of making digital images. The type of collection also played a major role in our determination of which digital images would be useful in the portal environment.

## REFERENCES

Chan, Winnie S. 2002. Cooperative Investment in Delivering Engineering Research Reports. In *Engineering Education in a Changing Economy, Proceedings of the Illinois/Indiana Sectional Conference*, April 11-12, Chicago, Illinois. Chicago: ASEE and Illinois Institute of Technology.

Fleischhauer, Carl. 1998. *Digital Formats for Content Reproductions.* http://memory. loc.gov/ammem/formats.html. Accessed May 6, 2002.

Mischo, William H. and Mary C. Schlembach. 1999. Web-based Access to Locally Developed Databases. *Library Computing* 18(1): 51-58.

Thompson, Larry A. 2001. Grey Literature in Engineering. *Science & Technology Libraries* 19(3/4): 57-73.

Zulu, Saul F.C. 1993. Towards Achieving Bibliographic Control of Unpublished Reports in Africa. *Libri* 43(2): 123-133.

# Challenges for Engineering Libraries: Supporting Research and Teaching in a Cross-Disciplinary Environment

Linda G. Ackerson

**SUMMARY.** The long-term relationship between engineering education, academic research, and the National Science Foundation has revolutionized the nature of engineering research and education. This paper focuses on engineers who work and teach in cross-disciplinary environments and use information from diverse disciplines. As disciplinary literature becomes increasingly fragmented and the volume of published information larger, engineers feel overwhelmed, trying to keep up with current research and teaching methods. These obstacles make it difficult for engineers to maintain an intellectual grasp of information, and challenge librarians to provide adequate physical access and manage collections. Two case studies are included to illustrate the problems of research and teaching in a cross-disciplinary environment. *[Article copies available for a fee from The Haworth Document Delivery Service: 1-800-HAWORTH. E-mail address: <docdelivery@haworthpress.com> Website: <http://www.HaworthPress. com> © 2001 by The Haworth Press, Inc. All rights reserved.]*

---

Linda G. Ackerson, MLS, is Assistant Engineering Librarian, Grainger Engineering Library Information Center, University of Illinois at Urbana-Champaign (E-mail: lackerso@uiuc.edu).

[Haworth co-indexing entry note]: "Challenges for Engineering Libraries: Supporting Research and Teaching in a Cross-Disciplinary Environment." Ackerson, Linda G. Co-published simultaneously in *Science & Technology Libraries* (The Haworth Information Press, an imprint of The Haworth Press, Inc.) Vol. 21, No. 1/2, 2001, pp. 43-52; and: *Information Practice in Science and Technology: Evolving Challenges and New Directions* (ed: Mary C. Schlembach) The Haworth Information Press, an imprint of The Haworth Press, Inc., 2001, pp. 43-52. Single or multiple copies of this article are available for a fee from The Haworth Document Delivery Service [1-800-HAWORTH, 9:00 a.m. - 5:00 p.m. (EST). E-mail address: docdelivery@haworth press.com].

10.1300/J122v21n01_05

**KEYWORDS.** Engineering education, cross-disciplinary research, cross-disciplinary teaching

## INTRODUCTION

The progression of engineers from apprentices in the "shop culture" to students in the "school culture" to researchers and teachers in a "cross-disciplinary culture" is a chronicle of the dynamic relationship between engineering education, academic research, and economic growth (Seely 1999). A particularly interesting part of the story involves the growth of interdisciplinary research and the role the National Science Foundation played in influencing the nature of engineering research and teaching on United States campuses. Today, engineering is a complex discipline dependent on the literature from multiple disciplines in order to advance. The majority of studies on cross-disciplinarity concern the research of scientists. This paper focuses on engineers who work in cross-disciplinary environments and includes two case studies to illustrate the influence on teaching, as well as research.

## HISTORY OF COOPERATIVE ENGINEERING EDUCATION AND RESEARCH

Traditionally, engineers were educated through a "learning by doing" approach by gaining skills in a machine shop or out in the field, until they had acquired enough skills to work on their own. Around 1880, individuals interested in engineering began enrolling in university courses, where the emphasis was still on the value of hands-on experience, but some science courses were included to enhance the development of practical skills (Seely 1999). In the 1930s, the Massachusetts Institute of Technology and Stanford University were two of the early leaders introducing the European model of education into United States engineering schools (Vest 1992-1993). European engineers recognized the value of using physics and mathematics as tools for solving problems and advocated a theoretical framework within which engineering problems could be examined.

World War II was a pivotal point for engineering, and in many respects, the nature of present-day research and engineering practice was shaped by the events of that era. Engineers had traditionally been employed as consultants by government and industry to handle real-world problems, such as designing dams and bridges, overseeing construction projects, and testing materials. When World War II began, there was a critical need for research on military projects involving computation, electronics, propulsion, materials, and other applications. To meet this need, the federal government offered substantial funding to universities to do basic research on technology. Engineers trained in engineering science had an advantage, because this type of research required knowledge of mathematics, chemistry, physics, and a more abstract approach to problem-solving.

Because of the successful cooperation between universities and government, the National Science Foundation (NSF) was created to continue support for research that would assure the United States would be military-ready in the advent of another war (Metzger and Zare 1999). The National Science Foundation played a major role in refashioning engineering education and research on university campuses. The idea of infusing scientific principles into engineering was adopted across the country. Engineering science became a mainstay in engineering education, and academic researchers trained their students to do basic research. The teaching of practical skills became less emphasized.

Over time, engineers involved in basic research began to divide and subdivide the areas in which they worked, partly as a strategy for retaining expertise in their areas and to keep up with the overwhelming volume of published information (Vest 1992-1993). This, too, defined the way in which engineering educators taught their students, who became experts in disciplines with an impressive depth of knowledge in that specialty. Years of focused study tended to produce engineers who were overspecialized and who lacked the skills and experience needed to work in a broader context and manage a complex system (Vest 1992-1993). In response, the National Science Foundation financially supported the creation of engineering research centers to encourage the synthesis of research across engineering disciplines. These units were free-standing programs, unattached to specific academic departments, and provided a place where engineers from many fields could work as a team, pooling their knowledge and sharing equipment, computing resources, and laboratories. Engineering research centers were meant to address long-range problems instead of focusing on immediate needs (Lewis 1987). The last available survey found that about half of the research conducted in these centers concerned materials, energy, and the environment, with lesser amounts devoted to manufacturing, communications, transportation, and biotechnology (Schowalter et al. 1995).

Graduate students had long been an important part of research on campuses that had strong research programs, but undergraduate students had played a lesser role (Morgan 1993).

In 1986, the National Science Board Task Committee on Undergraduate Science and Engineering Education commissioned a study of the present state of undergraduate education in the United States. Engineering was of particular concern, because the bachelors of science in engineering was often a terminal degree for engineers entering the work force. The findings of that study were published in the report *Undergraduate Science, Mathematics and Engineering Education* (often referred to as the Neal Report after the committee chair Homer A. Neal) (Neal 1986). The committee found that undergraduate engineering education failed to prepare students to make the transition from school to work. A primary problem involved the lack of hands-on experience. Too much of the education of engineers was based on textbooks, and the solution of contrived problems was substituted for involvement in real-life situations. The committee recognized that present-day problems demanded knowledge from several disciplines and that engineers needed to work together to solve them. The committee concluded that the best model for engineering education is a balance between engineering science and practice.

Three of the most serious problems identified by the committee were:

- Laboratory instruction was inadequate, in part because equipment and facilities were outdated.
- Faculty were unable to remain current in their areas due to heavy teaching loads or the inability to keep up with the excessive volume of research information published.
- The engineering curriculum did not adjust to constant changes, because it did not incorporate new teaching methods or advancements in engineering knowledge.

The National Science Foundation addressed these deficiencies by developing a series of programs (George et al. 1996):

- The Laboratory Development program and the Instructional Instrumentation and Equipment program whose main charge was to develop models of undergraduate laboratory instruction.
- The Faculty Professional Enhancement program which was designed to help engineering faculty experiment with innovative

teaching methods and introduce new material into coursework. (The Presidential Young Investigators program, which began in 1983, was primarily created to recruit and retain engineers in academic teaching and research careers (Palca 1990). The Faculty Professional Enhancement program, on the other hand, focused on improving instruction.)

- The Course and Curriculum Development program which funded the development of interdisciplinary courses and laboratories.
- The Research Experiences for Undergraduates program which allowed undergraduate students to participate directly in real-world research projects.

Significant changes were made in engineering curriculums to reflect the diversity of the engineering discipline. Many universities now offer courses on engineering ethics to explore the ties between technology and society and emphasize the engineer's responsibility of working within a social environment. Clarkson University offers a team-based design course that draws on the pre-requisites of physics, calculus, and chemistry and requires knowledge of hydraulics, environmental engineering, fluid mechanics, soil mechanics, engineering economics, and materials science to solve complex problems (Zander et al. 1995). Civil engineering students at Georgia Institute of Technology were given a unique opportunity to study the preparation of Atlanta for the 1996 Olympics (Chinowsky and Vanegas 1995). Students majoring in different subfields of civil engineering, such as construction and structural engineering, urban planning, and architecture, worked together to study the impact of adding factors such as mass transit, urban rehabilitation, and environmental impact to the planning process.

## ACHIEVING INTELLECTUAL ACCESS TO INFORMATION

Many terms are used to describe the use of literature from more than one discipline. Julie Thompson Klein devoted an entire chapter to the discussion of terminology (Klein 1990) and many authors have included varying descriptions of this concept in their own papers. However, there is no consensus about the definitions of these terms or their relationship to one another. The most frequently used words are inter-disciplinary, multi-disciplinary, trans-disciplinary, interdependent, and cross-disciplinary. For the sake of clarity, the term cross-disciplinary will be used in this paper as an umbrella concept that covers any type of research or literature searching methods that involve multiple disciplines. Two problems consistently identified in cross-disciplinary work are the fracture and overload of information (Palmer 1996). Both problems describe the challenges in maintaining an intellectual grasp of information.

The term fracture refers to the tendency for disciplinary literature to become fragmented as it is diffused into related disciplines and as new intersections are formed within a single discipline. It is difficult to know if all relevant information has been found, especially in a rapidly developing area, because so much of it resides on the periphery. Environmental engineering, for example, is highly cross-disciplinary because it was formed from the intersection of civil engineering, public health, ecology, and ethics. The term "environmental engineering" was first used in the 1960s to describe the interaction between academic programs for public health and engineering (Vesilind and Peirce 1982).

I interviewed Professor A, who works in a civil and environmental engineering department and does research in environmental engineering, specializing in membrane science. Professor A told me that he uses the literature of biology, chemistry, and physics as often as that of engineering, so his information needs span several disciplines. His favorite method of keeping up with current research is browsing, and he has a wide variety of tables-of-contents from both journals and books mailed to him each month. His second favorite method of searching the literature is following up on references. Although cross-disciplinary research is frequently considered to be at the cutting-edge, Professor A explained that because he uses the literature of traditional science disciplines, he often looks for older articles (defined as more than five years old) in search of seminal articles on a subject.

The term overload describes the situation of being unable to digest the volume of information being published from both core and peripheral areas. I also interviewed Professor B, who teaches engineering psychology, a cross-disciplinary field that resides in the intersection between three main areas: physiology and anatomy, experimental and social psychology, and physics and engineering. This course is required for majors in aviation engineering and industrial engineering and serves as the human factors course for other engineering students. It teaches the necessity of consulting many types of information to achieve effective designs. Measurements of anthropometric and biomechanical data can be found in military standards and technical reports, and studies on human memory and pattern recognition are published in the psychology literature. Also useful are observations of the effects of external stimulation such as noise and vibration on humans, most likely to appear as tabular data in handbooks.

Professor B told me that it was very difficult to find a suitable textbook that addressed the intersection of engineering and psychology, so it was important to him to be able to deposit diverse materials in one place. Putting supplemental readings in sociology, education, psychology, and engineering on reserve kept his students from having to purchase multiple textbooks. Regarding his own research, Professor B develops tools to measure the performance of air traffic controllers and studies the psychology of piloting aircraft. "I can't keep up," he told me. "I have a pile of papers [a foot high] that I've copied and can't get around to reading."

## OBSTACLES TO INTELLECTUAL ACCESS

Researchers have a difficult time searching for information across disciplines because the terminology is unfamiliar. Morecki, a researcher in the area of aircraft engineering and applied mechanics, appealed to his colleagues to address the standardization of terminology in inter-disciplinary areas, using the example of biomechanics (Morecki 1983). Biomechanics is a long-established area of anatomists, physiologists, and physicians, and the earliest terms used to describe concepts in this area were based on biology. When engineers began working in this area, they created new concepts in biomechanics by employing mathematical modeling and systems engineering, and inventing new terms such as motion study and rehabilitation engineering. It is easy for researchers to miss relevant research studies, even when using subject thesauri to identify related terms, because the irregular publication schedule of thesauri cannot always capture emerging trends in newly-developing subject areas. In addition, researchers are dependent on the quality of indexing. The number of access points assigned to documents must be sufficient to find a particular item among many similar ones. Otherwise, even the most carefully designed search strategy will not retrieve it (Swanson 1986).

Finding fragmented information may be obstructed by the use of bibliographic classification systems. The function of classification systems like Dewey Decimal and the Library of Congress is to organize the entire universe of information by assigning it into classes and sub-classes that largely follow the traditional disciplinary structure (Beghtol 1998). When areas of knowledge merge or overlap with other areas, a strict classification structure cannot always accommodate these connections. The initial strategy is to shoehorn the information into a place that provides a "good enough" fit. The assignment of unique subject headings often lags far behind the evolution of a new subject area, and they are finally assigned only when it become obvious that the information in an area will exceed the category in which it was first allocated. For example, a significant amount of literature on the environment was produced during the late 1980 and early 1990s. However, the Library of Congress subject heading "environmental sciences" for works on the "composite of physical, biological, and social sciences concerned with the conditions of the environment and their effects" was not created until 1992 (OCLC authority record number 3274635).

## ACHIEVING PHYSICAL ACCESS TO INFORMATION

A major concern for cross-disciplinary researchers and educators is the physical availability of materials, whether in print or electronic format. Cross-disciplinary research and teaching are best served by strong disciplinary collections of primary sources on which they can draw. In our interview, Professor A mentioned that specialized journals like *Journal of Membrane Science* are important for his work, but society journals such as the *AIChE Journal* and the journals and conference proceedings of the American Water Works Association are just as important.

Along with sufficient quantity, engineers need access to a diversity of information sources to accommodate cross-disciplinary research and teaching. For example, students enrolled in engineering ethics courses need access to a wide variety of sources to understand the concept of working within a social environment. Searching for information on the Three Mile Island catastrophe will lead students to books and articles on the subject, but they should also have access to the report produced by the Nuclear Regulatory Agency about this event (Hagler 1997). Professor B noted that students in human factors courses need to use a combination of data compilations, organizational standards, and published papers to gather sufficient information for a design project.

## OBSTACLES TO PHYSICAL ACCESS

Collection management decisions affect the quality of the collection for cross-disciplinary research and teaching (Searing 1996). Effective collection development often depends on the degree of collaboration among subject specialists. Cross-disciplinary areas may lack the assessment tools and subject bibliographies by which collections are measured, because the area is too new or too specialized to have been covered. Therefore, it is important to make sure that newly-emerging areas are added to the approval profile and that appropriate societal publications are consistently monitored. Cross-disciplinary areas like biotechnology, bioinformatics, and tissue engineering now produce a significant amount of literature, but they do not fit neatly into the disciplinary categories on which fund assignments are based, and selectors may be hesitant to adopt new areas in lean budget years.

A long-running debate in the library science literature concerns the most effective organization of libraries. Should campus libraries be centralized or decentralized, and which arrangement is better for cross-disciplinary work? With the growing number of electronic full-text resources and article databases accessible from offices, labs, and homes; the physical arrangement of libraries is less important to faculty and students today.

## CONCLUSION

In the past, engineering education was based on service and technical experience. The current revolution in engineering is the transition from a service-based economy to a knowledge-based economy. Greater numbers of students are enrolling in information technology courses, where emphasis is placed on analyzing and synthesizing knowledge along with the development of technical skills (Chang 2001). Now information technology skills are fast becoming as important for engineers as experience in applying human factors data. In rapidly evolving cross-disciplinary areas, keeping up with current literature present challenges for librarians as well as for engineers.

## REFERENCES

Beghtol, Clare. 1998. Knowledge Domains: Multidisciplinarity and Bibliographic Classification Systems. *Knowledge Organization* 25 (1-2): 1-12.
Chang, David C. 2001. Engineering Education: Where Do We Go from Here? *IEEE Antennas and Propagation Magazine* 43 (December): 114-116.

Chinowsky, Paul S. and Vanegas, Jorge A. 1995. Facilitating Interdisciplinary Civil Engineering Education Through a Living Laboratory. In *Investing in the Future ASEE Annual Conference Proceedings*. Washington, DC: American Society for Engineering Education.

George, Melvin D. et al. 1996. Shaping the Future: New Expectations for Undergraduate Education in Science, Mathematics, Engineering, and Technology. *Report prepared for National Science Foundation Directorate for Education and Human Resources* (http://www.ehr.nsf.gov).

Hagler, M.O. et al. 1997. Editorial. *IEEE Transactions on Education* 40 (November): 289-290.

Klein, Julie Thompson. 1990. *Interdisciplinarity: History, Theory & Practice* Detroit: Wayne State University Press.

Lewis, Courtland. 1987. Interdisciplinary Engineering Research: A Case Study. *Engineering Education* 78 (October): 19-22.

Metzger, Norman and Zare, Richard N. 1999. Interdisciplinary Research: From Belief to Reality. *Science* 283 (January): 642-643.

Morecki, Adam. 1983. On the Study of the Standardization of Terminology in Interdisciplinary Sciences. *Mechanism and Machine Theory* 18 (April): 225-227.

Morgan, Robert P. et al. 1993. Engineering Research in U.S. Universities: How University Research Directors See It. In *Frontiers in Education ASEE Annual Conference Proceedings*. Washington, DC: American Society for Engineering Education.

Neal, Homer A. 1986. Undergraduate Science, Mathematics and Engineering Education. Report prepared for the National Science Board.

Palca, Joseph. 1990. Young Investigators at Risk. *Science* 249 (July 27): 351-353.

Palmer, Carole L. 1996. Information Work at the Boundaries of Science: Linking Library Services to Research Practices. *Library Trends* 45 (Fall): 165-191.

Searing, Susan E. 1996. Meeting the Information Needs of Interdisciplinary Scholars: Issues for Administrators of Large University Libraries. *Library Trends* 45 (Fall): 315-342.

Schowalter, William R. et al. 1995. Forces Shaping the U.S. Academic Engineering Research Enterprise. *Report prepared for the National Science Foundation.*

Seely, Bruce E. 1999. The Other Re-engineering of Engineering Education, 1900-1965 *Journal of Engineering Education* 88 (July): 285-294.

Swanson, Don R. 1986. Undiscovered Public Knowledge. *Library Quarterly* 56 (April): 103-118.

Vesilind, P. Aarne and Peirce, J. Jeffrey. 1982. *Environmental Engineering*. Ann Arbor: Ann Arbor Science.

Vest, Charles M. Report of the President for the Academic Year 1992-1993. *President Charles M. Vest's annual report to the MIT community*, October 1993. Available at: http://web.mit.edu/president/communications/rpt92-93.html.

Zander, A.K. et al. 1995. Advantages and Organization of Interdisciplinary Design Projects. *Investing in the Future ASEE Annual Conference Proceedings*. Washington, DC: American Society for Engineering Education.

# Information-Seeking Behavior
# of Academic Meteorologists
# and the Role of Information Specialists

Julie Hallmark

**SUMMARY.** In lengthy interviews of 43 meteorologists at two universities and a research center, the author investigated their methods of seeking information needed for their research, teaching, and current awareness. The primary goals of the study were to determine problems and challenges encountered by these scientists and to develop a profile of their information needs. Of particular interest were the effects of the Internet on their information-seeking behavior, particularly their access and retrieval of electronic journal articles and data. Suggestions for academic science librarians were developed through another series of interviews with successful information professionals. More than ever before, scientific information specialists must make clear their new roles in facilitating the scientific endeavor, especially the constant effort and vigilance required in developing and maintaining library Web sites, so convenient and indispensable to their clientele. *[Article copies available for a fee from The Haworth Document Delivery Service: 1-800-HAWORTH. E-mail address: <docdelivery@haworthpress.com> Website: <http://www.Haworth Press.com> © 2001 by The Haworth Press, Inc. All rights reserved.]*

Julie Hallmark, MLS, PhD (Library and Information Science), is Professor, Graduate School of Library and Information Science, The University of Texas at Austin, Austin, TX.

[Haworth co-indexing entry note]: "Information-Seeking Behavior of Academic Meteorologists and the Role of Information Specialists." Hallmark, Julie. Co-published simultaneously in *Science & Technology Libraries* (The Haworth Information Press, an imprint of The Haworth Press, Inc.) Vol. 21, No. 1/2, 2001, pp. 53-64; and: *Information Practice in Science and Technology: Evolving Challenges and New Directions* (ed: Mary C. Schlembach) The Haworth Information Press, an imprint of The Haworth Press, Inc., 2001, pp. 53-64. Single or multiple copies of this article are available for a fee from The Haworth Document Delivery Service [1-800-HAWORTH, 9:00 a.m. - 5:00 p.m. (EST). E-mail address: docdelivery@haworthpress.com].

10.1300/J122v21n01_06

**KEYWORDS.** Meteorology, scientific communication, information-seeking behavior

## INTRODUCTION

In the present environment of rapid email communication and increasing numbers of electronic and web-based resources, librarians seek new and more effective roles as productive, active contributors to the scientific endeavor. To accomplish this goal it is critical that information specialists know as much about their clientele as possible; at the same time, scientists need to be aware of support and services offered by their librarians. This research investigates user needs and information-seeking behavior of academic meteorologists with the goal of gaining greater insight into challenges and problems faced by this user community.

Although fewer users physically visit the academic science library (with forecasts of even fewer in the future), their information needs are often complex. The goals of the present research were to investigate and clarify these information needs, specifically:

- to determine *problems* encountered by meteorologists as they seek and acquire the data and other information needed for their research
- to develop a *profile of the information needs* of the academic meteorologist
- to summarize suggestions for *academic science librarians* that emerged in interviews with successful professionals.

Whether acquiring expensive databases, electronic journals, or the new edition of an encyclopedia, the meteorology librarian must deal with licensing issues and restrictions, budget shortfalls, Web maintenance, and competing demands on library resources from clientele. The broad interdisciplinary range of interests of atmospheric scientists adds to the challenge of providing high quality information services.

Another significant factor in today's information equation is the scientific journal publisher, some of whom continue to establish more and more restrictive policies of use and access while increasing subscription costs. Clearly, effective and continuing communication between scientists and their information specialists is more critical than ever, as they develop strategies to deal with such policies.

## BACKGROUND

Information needs and information-seeking behavior of scientists have been of interest to librarians for many years. The last decade, particularly, has seen rapid evolution to new electronic information environments that have revolutionized the ways scientists work (Crawford, Hurd, and Weller 1996). By way of illustration, in 1991 most faculty members in the health sciences at the University of Illinois at Chicago used print indexes as their preferred access to the literature even after online versions became available. By 1995, 68% of the medical faculty accessed *Medline* electronically (Curtis, Weller, and Hurd 1997). Since 1998 *Index Medicus* has not been available in print format, and researchers around the world use only the electronic *Medline*.

A productive line of investigation in recent years is based on the "grounded theory" of Glaser and Strauss (1967), a form of naturalistic inquiry. This approach models information-seeking behavior by establishing generic characteristics of information-seeking patterns through in-depth interviews. Ellis (1993) used the grounded theory approach to study academic researchers; Ellis, Cox, and Hall (1993) used this theory to investigate researchers in the physical and social sciences; and Ellis and Haugan (1997) used it to study engineers and scientists in industry.

In contrast to naturalistic inquiry that identifies the changing patterns of information needs and behavior throughout the phases of a project, the present study employed one-time, open-ended interviews along with a brief quantitative citation analysis, resulting in a "snapshot" of information-seeking behavior. Particularly relevant were former studies that focused on specific discipline(s) through interviews and questionnaires, for example, Hallmark (1994)–physicists, mathematicians, chemists, biologists, geologists; Rolinson, Al-Shanbari, and Meadows (1996)–biological researchers; Brown (1999)–astronomers, chemists, mathematicians, and physicists; Curtis, Weller, and Hurd (1997)–health sciences faculty; Hurd and Weller (1997)–academic chemists; and Hurd, Blecic, and Vishwanatham (1999)–molecular biologists. These studies, taken as a whole, reflect several aspects of a proposed model of scientific communication for 2020, one that is reminiscent of the physicists' long-standing *modus operandi* in their preprint culture (Hurd 2000).

## METHODOLOGY

The method used for gathering data from scientists was a series of individual interviews with 43 faculty, administrators, and graduate students in atmospheric sciences and meteorology at Texas A&M, Oklahoma University, and the National Severe Storms Center in Norman, Oklahoma. These sites were selected because of their excellent reputations, locations convenient to the researcher, and different modes of providing library and information service.

Interviews began with the researcher's requesting a reprint of a recent journal article from the scientist. Then the scientist was asked to comment briefly on the access and retrieval of one or two of the citations in the bibliography of the article. Further conversation focused on the following topics:

- significant effects of the Internet on access and retrieval of information resources
- challenges in obtaining information relevant to their research such as conference proceedings, technical reports, data, and software.

The scientists explained how they kept current in their field and how they used (or did not use) the library. They candidly discussed issues such as attracting and retaining graduate students, funding academic programs in times of shrinking support from the university, and acquiring data needed for their research.

In addition, a series of open-ended interviews with successful science information professionals elicited suggestions and advice for their colleagues regarding library services in a series of "what works for me" conversations.

## RESULTS

A well-funded, focused science branch library on a university campus, staffed by information professionals and dedicated to services for relatively few users, is ideal. Frequently found in academic departments of chemistry and geology, such information services are very much appreciated by their users who will "fight to the death" any suggestion that "their" library be integrated into a more general facility. But even in such ideal environments, constant marketing and publicity are critical to the success of the information service.

There has always been an obvious relation between strong financial support of a high-quality, professional information service and user satisfaction and awareness of the service. Several of the scientists interviewed in the present study commented on the need for better financial support of their local information services and resented, in some instances, the apparent lack of concern for their needs on the part of the higher administration.

As with other scientists, journal article access and retrieval is a critical information need of meteorologists. Based on a sample of citations from their articles, two-thirds of the references cited in these published papers come to their attention through *personal contacts* (suggestions from colleagues) and *references in the published journal literature* rather than through database searching. Thus, this group of meteorologists, along with scientists in other disciplines, tends to use the easiest and quickest methods of locating relevant journal articles, not necessarily the best. Electronic journals are an enormous asset to the researcher, but only if subscribed to by the institution, obviously. Access to e-journals may be the latest indicator of information "haves" and "have-nots."

In the eyes of most of those interviewed, the most critical roles of the library are that of monograph and journal repository, plus the acquisition of items not owned locally through interlibrary loan or document delivery. Relatively few scientists interviewed could be classified as sophisticated library users, i.e., people who are clearly aware of the range of services offered and who would readily ask the librarian for assistance beyond simple article retrieval, verification of an address or citation, or retrieval of a specific datum from a handbook. On the other hand, several recalled instances of very high-level service they had experienced from their information specialists. The same old question continues to arise: "How can we make our clientele aware of the broad range and complexity of services they can expect from their librarian?"

## *Information Needs and Problems*

As represented through these interviews the academic atmospheric scientist:

- feels a strong sense of lack of time as he or she balances teaching, research, graduate student support and collaboration, budgetary efforts to keep the department afloat, and grant applications. In describing his need for speed, one researcher commented, "I have to work quickly; often I just read the abstract instead of going to the library."

- depends heavily on Internet resources for text and data; uses electronic formats for data and new journal articles almost exclusively.
- typically uses the American Meteorological Society (AMS), Meteorological and Geoastrophysical Abstracts (MGA), American Geophysical Union (AGU), or Web of Science from the Institute of Scientific Information (ISI) websites to begin a search.
- is a heavy user of technical reports from National Aeronautics and Space Administration (NASA), National Oceanic & Atmospheric Administration (NOAA), and other government agencies.
- visits the physical library much less frequently than in the past, substituting library Web pages as a major resource.
- may have difficulties obtaining data and software described in published research. The complex models in meteorology are heavily data-driven, and atmospheric scientists deal with challenges and frustrations surrounding the acquisition, storage, and use of software and data sets. A graduate student described his research using NASA's Tropical Rainfall Measuring Mission (TRMM) data, "designed to monitor and study tropical rainfall and the associated release of energy that helps to power the global atmospheric circulation shaping both weather and climate around the globe." Another researcher commented, "For weather forecasting we get data sets from the National Weather Service–1/2 gigabyte per day for our medium range forecast model (MRF) at no cost. Radar and satellite data can be 2-3 gigabytes per day. Storage is a problem, as data sets get bigger and bigger. We have a 50 terabyte system shared with the College of Science."

When asked about acquiring data, one professor responded, "Ah, data. That's the problem." He went on to describe issues that may arise with regard to data: the cost, dealing with various government agencies, and sometimes simply identifying the existence of a particular data set. Data from commercial sources are expensive, for example, lightning data from Global Atmospherics Inc. that include all lightning strikes in the lower 48 states for the last two decades with data for each lightning strike: whether positive or negative, peak current, number of strokes, etc.

Concerning software, another researcher commented:

> It's hard to get hold of software–it's written as part of a project; after the project is over, the software is useless and there's no

documentation! We need a software depository. (You can't publish software in a paper.) We often depend on the author to send the software and that occasionally works. Frequently we hire a graduate student who leaves after the project is over. One solution would be to "cement" software and data tapes somehow. In our own department a graduate student left; we couldn't reproduce the results and had to redo everything. Right in the same department where the work was done! Maybe we should store software info on the AMS website. Software drifts around like jetsam.

- is often unaware of standard library services. Filling straightforward interlibrary loan requests for monographs or provision of older, foreign journal articles seemed amazing to some of those interviewed.
- must constantly and unrelentingly seek grants and contracts to support his students and laboratories. "An academic department operates just like a business; university grant money is very scarce. We must attract good students. Formula funding is weighted toward doctoral and science students."
- enjoys browsing as time permits. Some admitted a feeling of guilt because they don't come to the library so much anymore. "I'm getting lazy; I feel like I am missing stuff." "Hunting, clicking, and scrolling is not the same as browsing." "About the only time I actually go to the library is to look at the new books." "I don't have time to go the library anymore; I just send my graduate students to find stuff."
- is pleased to talk with someone who is interested in their frustrations and concerns and who seeks their opinions regarding solutions.
- when asked what library services he/she would most like to have, invariably mentions photocopied articles. "Everyone wants articles copied, whether they come from print or online resources, unless they are simply preparing a manuscript and inserting citations," explained one scientist. One assistant professor was positively eloquent when describing the "librarians" at the California Institute of Technology who photocopied for him. Might the library consider having a student assistant copy articles for users, charging costs to the scientist's account? Two scientists, by the way, pointed out that they prefer to request reprints from the authors since illustrations are in color.
- has positive, friendly feelings toward libraries and librarians but visits the library much less frequently.

- adheres to the "law of proximity" that has always influenced our users' information-seeking behavior and continues to be a major factor in library use. All science librarians know the uproar that occurs if the library is moved to another floor, or, worse, to another building. Now, interestingly enough, it seems that the relationship between the physical location of the library and the use it receives is applicable to the Web environment.

In universities, science faculty members often use the department Web site as their home page. An informal study at The University of Texas at Austin showed that the two branch libraries with links off the departmental Web sites had many more hits than the other three libraries, thus the law of proximity:

> The use of the library varies inversely with the *distance* the user is physically from the library (down to a level of linear feet) or the *number of clicks* necessary to get to a Web page. (Flaxbart 2002)

## Suggestions for Academic Science Librarians

As environments in science libraries change, how can science information specialists remain critical to their clientele? For years members of the Special Libraries Association have emphasized their role as facilitator, guide, and navigator for their users. The need for such individuals emerged yet again in the present research. For example, two users who were interviewed had access to *Meteorological and Geoastrophysical Abstracts* but did not realize it; another had access but had never heard of the database–after trying it out in his office, he was surprised and pleased.

The following suggestions with regard to the present environment in university science libraries come from conversations with academic branch librarians representing a variety of institutions and disciplines (chemistry, biology, geology, engineering and physics):

- University science librarians have new opportunities for integrating collections. One can "mix and match" journal titles now that physical location isn't an issue. For example, the same electronic journal can be listed on two branch libraries' Web sites.
- Information specialists can discover new ways of providing services and effective public relations. Scientists need to be told what the library is providing and how it adds value. Flaxbart informs his

clients of the enormous effort necessary to acquire and maintain electronic resources–selection, licensing, cataloging and making links, and the ever-present trouble shooting. "An electronic structure isn't permanent; it's fragile and requires constant negotiation with the provider. For example, a single hiccup in the system can remove all the journals of a publisher," he points out. Of course, the library staff must continue to select, acquire, and catalog monographs.

- Offer an appealing new service and don't stop to ask "Will I be overwhelmed if this takes off?" Being overwhelmed just provides proof that you need more help and more money. Tailor new services to individual faculty members, starting with the people who are most influential or who are big library users already. One engineering librarian commented, "You can be a wizard in your users' eyes."
- Keep users apprised and up-to-date regarding services and problems with publishers. Protests from scientists can be effective, as was the case some years ago when the Geological Society of America made unpopular changes in the format of the *GSA Bulletin*.
- Attend your users' brown bag talks and seminars. My experiences when visiting campuses were most informative. Texas A&M researchers described an ongoing project on smog in Houston–what causes smog, how it is monitored, how it affects surrounding areas. At the University of Oklahoma a member of the meteorology team that was based in Sydney for the Olympics described the challenges and the methodologies of forecasting weather events that might affect the track, beach volleyball, and other venues using the sophisticated "Australian Integrated Forecast System." Such informal gatherings offer a great opportunity for the librarian to interact with users, whether on a university campus, in a government agency, or in a company.
- Offer a brown bag of your own, for example, to demonstrate a new product–the "Electronic Resource of the Month." Graduate students will come, for the doughnuts if nothing else. Ask for a slot from time to time in the departmental Friday afternoon seminar schedule.
- Be everywhere your users are. One branch librarian commented that he got two requests for help at the annual departmental picnic recently.
- Seek out new users. They may be totally uninformed about the library and its services, even though they participated in an orienta-

tion or received email or a brochure describing library services upon joining the organization. Scheduling a meeting with every user in his or her office once a year to catch up on their research and learn about new directions, interests, and current problems is effective.

- Encourage browsing in the library. A pleasant "New Books Area" is definitely a major draw for scientific clientele. Students and faculty say they would like a more Borders/Barnes & Noble feel to the library including comfortable couches, arm chairs, coffee and food. "Why not? We never could enforce the food embargo anyway," admits one branch librarian.
- Find new ways of providing services and improving public relations. Scientists need to know what their libraries are providing and how they add value. One librarian, for example, volunteered to do a bibliographic search for new grant proposals.
- Rethink library statistics. Many science librarians keep usage statistics on all pages in their Web site. They can determine which pages on their site get the most traffic and estimate the number of individuals that view a certain page. Are we seeing a modified reincarnation of counting the number of reference books left on tables at the end of the day?
- Email articles to users without specific requests; know their interests well enough to spot likely candidates. Route new books to users based on your knowledge of their research.
- Make appointments from time to time with your users for 30-45 minutes to discuss their research, problems, frustrations, and suggestions. They will appreciate your interest.

Science-technology librarians could easily add more examples to these ideas, as they continue to develop new, innovative approaches to the provision of information services.

## CONCLUSIONS

This investigation presents a snapshot of academic meteorologists and their information needs at the beginning of the twenty-first century. Confirming similar studies in other disciplines, these scientists are often unaware of services available to them and may hold surprising misconceptions. The journal article, whether print or electronic, continues to be their ultimate textual resource. Their research is heavily data-driven, and the successful meteorological librarian must become familiar with numerous acronyms and government agencies.

These academic researchers are overworked and face conflicting demands on their time from graduate student supervision and support, class preparations, grant applications, production of refereed journal articles, tenure decisions, external political maneuvering over matters such as competition for physical space, and a host of other issues. It is not surprising that they feel they do not have enough time for sophisticated information-gathering.

Clearly, meteorology information specialists, as well as those in other science-technology disciplines, are thriving. Although the physical collection shrinks daily, new opportunities for offering impressive information services to clientele through library Web sites seem unlimited. Librarians in this field, as in others, continue to be a critical part of the scientific endeavor.

## REFERENCES

American Geophysical Union. 2002. Retrieved February 20, 2002, from http://www.agu.org/.

American Meteorological Society. 2002. Retrieved February 20, 2002, from http://www.ametsoc.org/AMS/.

Brown, C. M. 1999. Information seeking behavior of scientists in the electronic information age: Astronomers, chemists, mathematicians, and physicists. *Journal of the American Society for Information Science* 50:929-943.

Crawford, S. W., Hurd, J. M. and Weller, A. C. 1996. *From print to electronic: The transformation of scientific communication.* Medford, NJ: American Society for Information Science.

Curtis, K. L., Weller, A. C., and Hurd, J. M. 1997. Information-seeking behavior of health sciences faculty: The impact of new information technologies. *Bulletin of the Medical Library Association,* 85:402-410.

Ellis, D. 1993. Modeling the information-seeking patterns of academic researchers: A grounded theory approach. *Library Quarterly* 63:469-486.

Ellis, D., Cox D., and Hall, K. 1993. A comparison of the information seeking patterns of researchers in the physical and social sciences. *Journal of Documentation* 49:356-369.

Ellis, D. and Haugan, M. 1997. Modeling the information seeking patterns of engineers and research scientists in an industrial environment. *Journal of Documentation* 53:384-403.

Flaxbart, D. 2002. [Conversations] The University of Texas at Austin.

Glaser, B. G. and Strauss, A. L. 1967. *The discovery of grounded theory: Strategies for qualitative research.* New York: Aldine.

Global Atmospherics Inc. 2002. Retrieved February 20, 2002, from http://www. lightningstorm.com/.

Hallmark, J. 1994. Scientists' access and retrieval of references cited in their recent journal articles. *College and Research Libraries* 55:199-209.

Hurd, J. M. 2000. The transformation of scientific communication: A model for 2020. *Journal of the American Society for Information Science*, 51:1279-1283.

Hurd, J. M., Blecic, D. D., and Vishwanatham, R. 1999. Information use by molecular biologists: Implications for library collections and services. *College and Research Libraries* 60(1):31-43.

Hurd, J. M. and Weller, A. C. 1997. From print to electronic: The adoption of information technology by academic chemists. *Science & Technology Libraries* 16(3/4): 147-70.

Institute for Scientific Information. 2002. Retrieved February 20, 2002, from http:// www.isinet.com/isi/.

Meteorological and Geoastrophysical Abstracts (MGA). 2002. Retrieved February 20, 2002, from http://www.csa.com/.

Rolinson, J., Al-Shanbari, H., and Meadows, A. J. 1996. "Information usage by biological researchers," *Journal of Information Science*, 22(1): 47-53.

Tropical Rainfall Measuring Mission. 2002. Retrieved February 20, 2002, from http://trmm.gsfc.nasa.gov/.

U.S. Department of Energy. *DOE Reports Bibliographic Database.* 2002. Retrieved February 20, 2002, from http://www.osti.gov/html/dra/dra.html.

World Meteorological Organization. 2002. Retrieved February 20, 2002, from http://www.wmo.ch/indexflash.html.

# Geology Librarianship:
# Current Trends and Challenges

Lura E. Joseph

**SUMMARY.** Current trends and challenges in geology librarianship are influenced by driving forces such as the exponential increase in knowledge, the interdisciplinary nature of sciences, specialization of disciplines, space and budget constraints, and user expectations. The current trend toward a digital environment affects all areas of librarianship including collection development, current awareness, reference, instruction, preservation, dissemination of information, legislation and copyright law. Although the future role of librarianship may appear murky, science librarians will almost certainly continue to function as information specialists who reach across disciplinary boundaries to find related information. They will continue to prepare users for a lifetime of learning, work to standardize the various electronic resources, and ensure that knowledge is preserved in some format. This article provides an overview of the current state of geology librarianship. *[Article copies available for a fee from The Haworth Document Delivery Service: 1-800-HAWORTH. E-mail address: <docdelivery@haworthpress. com> Website: <http://www.HaworthPress.com> © 2001 by The Haworth Press, Inc. All rights reserved.]*

Lura E. Joseph is Geology and Digital Projects Librarian and Assistant Professor, Library Administration, University of Illinois, Urbana-Champaign (E-mail: luraj@uiuc. edu). She received her MLIS from the University of Oklahoma. She also received master's degrees in Geology and Psychology and a BA in Anthropology.

[Haworth co-indexing entry note]: "Geology Librarianship: Current Trends and Challenges." Joseph, Lura E. Co-published simultaneously in *Science & Technology Libraries* (The Haworth Information Press, an imprint of The Haworth Press, Inc.) Vol. 21, No. 1/2, 2001, pp. 65-85; and: *Information Practice in Science and Technology: Evolving Challenges and New Directions* (ed: Mary C. Schlembach) The Haworth Information Press, an imprint of The Haworth Press, Inc., 2001, pp. 65-85. Single or multiple copies of this article are available for a fee from The Haworth Document Delivery Service [1-800-HAWORTH, 9:00 a.m. - 5:00 p.m. (EST). E-mail address: docdelivery@haworthpress.com].

10.1300/J122v21n01_07

**KEYWORDS.** Trends, challenges, geology, librarianship, digital format, electronic format

## *INTRODUCTION*

In 1945, Vannevar Bush wrote an article titled "As we may think" (Bush, 1945). In it, he commented on an exponential growth in research information, specialization, and the daunting task of bridging between disciplines. Bush also predicted the invention of a device he called "the memex." This would be a desktop tool that would store huge volumes of information in all formats, the contents of which could be easily searched and retrieved. The user could link information and build multiple "trails" which could be recalled in the future for personal and shared use. A researcher could have all the relevant literature in his laboratory with trails and side trails to information.

J. C. R. Licklider was another individual with uncanny foresight. Licklider, possibly one of the most influential individuals in the history of computer science, was responsible for setting the funding priorities that led to the creation of the Internet (Waldrop 2001). In 1968, Licklider and Taylor described computer networks and predicted on-line interactive communities (Licklider and Taylor 1968). Licklider had a vision of networked computing in which large numbers of people would be connected by a global network. Licklider's book *Libraries of the Future* was written in 1965, well before the advent of the online catalog. In the introduction, he stated:

> We need to substitute for the book a device that will make it easy to transmit information without transporting material, and that will not only present information to people but also process it for them, following procedures they specify, apply, monitor, and, if necessary, revise and reapply. To provide those services, a meld of library and computer is evidently required. (Licklider 1965)

Licklider went on to describe some of the computer techniques and procedures that could be used in creating "an automated card catalogue," and the use of Boolean functions as retrieval terms. Many of Licklider's 1965 predictions about libraries in the year 2000 had become reality before his death in 1990.

It is amazing how Bush's 1945 vision has come to pass with the modern personal computer, the Web, and electronic indexes, journals and books; and Licklider's predictions of the Internet and libraries have become a reality. A wealth of information is available at our fingertips. However the challenges accompanying an exponential increase in knowledge and increased specialization remain undiminished, indeed compounded. This exponential increase in knowledge is hardly new, however the responses to those challenges are changing.

## TRENDS AND CHALLENGES

### Driving Forces

Most of the trends and challenges related to geology librarianship are similar to those of science librarianship in general. In addition to the exponential increase in knowledge and the digital revolution, there are numerous additional driving forces.

Most of the sciences are becoming ever more interdisciplinary. Geology has always been interdisciplinary in nature, linking fields such as biology, chemistry, physics, physical anthropology and archaeology, and astronomy. Take for example the topic of "global climate change" which incorporates the study of pollen, diatoms, beetles, ice core contents, ocean thermohaline circulation, gas hydrates, mountain building, volcanic eruptions, Milankovitch cycles, and human impacts, to name a few. All of these factors may interact, and keeping up with the information being generated in all the related fields is a monumental task for researchers. Another driving force is the increased specialization within the various disciplines. This specialization fragments the disciplines, making it more likely to miss pertinent information in related areas.

Budget constraints are an ever-present challenge to most libraries, and continue to be a driving force. The prices of publications continue to soar. According to Association of Research Libraries (ARL) statistics, serials expenditures for ARL libraries have increased 210% from 1986 to 2001 (http://www.arl.org/stats/arlstat/graphs/2001/2001t4.html). Even the larger institutions have cancelled many or most of their duplicate journal titles. The fat is gone, and cuts are into the bone. In many cases larger research institutions are charged much higher rates for electronic resources, reducing their buying power. For example, the year 2003 rates that American Geophysical Union (AGU) charges Category 1 institutions (2 year Associate Degrees) versus Category 5 institutions (greater than 50 Doctorate graduates per year) for electronic access to *Journal of Geophysical Research* are $1,138 for Category 1 versus $13,444 for Category 5 (http://www.agu.org/pubs/Institution_ Rates_2003.pdf).

Space limitations are another driving force. The problem of ever expanding collections is not new, but some of the related choices are new. In addition to collection weeding or building high density storage facilities, there is the potential of relying on electronic archival collections. Another driving force is the need for preservation. Again, while the problem is not new, there are the new choices of electronic format.

User expectations are an additional challenge. As patrons adjust to the new technologies, they begin to expect everything to be available in digital format, accessible at any time, from any computer. They are increasingly impatient with lack of standardization, and with technical glitches. A large-scale survey at the University of Washington indicates that faculty and graduate students in science/engineering and health sciences are more likely to access library resources remotely rather than onsite, place desktop delivery as the highest priority for library service, and place a higher value on journals (both electronic and print) than on books (Hiller 2002).

### Digital Choices: Trend Toward Digital Formats

At the year 2000 Geological Society of America/Geoscience Information Society annual meeting, Regan gave an overview of the trends and future of the "intersection of geoscience, information technology, and society" and used the term "geoinformatics" (Regan 2000). Regan emphasized the transition from libraries as "physical storehouses of paper texts to electronic gateways serving virtual communities of users." As noted by Regan, when faced with format choices, there is a trend toward the digital format. On the other hand, since documents on the Web may disappear without warning, some librarians continue to print pertinent materials from the Web and add them to their print collections (Manson 2001). Digital choices include e-journals, e-books, and geographic data in electronic format.

### E-Journals

There are a number of obvious advantages to digital formats over print journals. They can save space and binding funds if the print version is cancelled. They potentially can be accessed from any place and at any time, by multiple users. There is normally no deterioration in copy quality. This is especially important in geology and other disciplines that frequently use photographs, micrographs, and color illustrations. In addition, access to current journal issues is often more timely.

There are some potential disadvantages to e-journals. Since license agreements typically restrict access to certain IP ranges, complex authentication procedures are sometimes necessary if they are to be accessed outside of the library. In addition, the cost of some e-journals for institutions is currently prohibitive. As the availability of e-journals increases, the ability to afford the material becomes ever more challenging (Duranceau, 2001).

Currently, the confusion generated by the myriad ways to access and operate the e-journals and search engines is a challenge to most users. Electronic indexes have been in existence for enough time that standardization should be emerging, however much is left to be done in this area. E-journals are in even more disarray. Currently e-journals may be accessed via publishers, aggregators, and indexing services, each with its own system. Some of the systems offered by publishers work very well, for example, ScienceDirect from Elsevier (http://www.sciencedirect. com). The journals can be browsed, or collectively searched. Some systems restrict searches only to journals owned by the library, while others offer universal searches and the option to purchase non-owned articles. Aggregators offer "packages" of electronic journals from a variety of publishers, and publishers are also offering their own electronic journals as package deals. American Association of Petroleum Geologists (AAPG), Geological Society of America (GSA), and the Society of Exploration Geophysicists (SEG) are cooperating to create an electronic journal aggregate, and other professional societies have expressed an interest in joining the effort (Noga 2002).

These groupings offer a partial solution to standardization of operation. However, some of the results from aggregating services are disappointing. There are problems of delay in adding titles, some issues or years may be missing, and titles may disappear from the collection overnight. Output options may also vary, with services offering a combination of HTML, PDF, and bit-mapped files. Some include graphics and tables; some do not. Graphics and tables may not always be important for undergraduates, but they are essential for graduate students and researchers in disciplines such as geology.

Some libraries are creating Web tools to simplify access to e-journals. The University of Illinois at Urbana-Champaign offers an example of an e-resource registry (Chan 2002). In this database-driven system, lists of e-journals are generated by the library, for example: Geology Journals (http://www.library.uiuc.edu/gex/journals.html). Another example is the e-resource index created by the North Dakota State University Libraries to simplify access to e-journals. E-journals are indexed by title, subject, and department. If access to a particular title is not allowed from a remote site, the title will not appear in the generated list. This resource can only be accessed by valid user logon. Maintaining these lists can be problematic. Access to particular e-journals and licensing arrangements may come and go, or the base URL may change. Access problems are sometimes known only if reported by users, and tracking down the source of the problem can be very frustrating and time consuming, as well as involving a number of individuals. Access via online library catalogs will be discussed later.

These in-house tools help simplify access to electronic journals, but do not address the lack of standardization of operation of the various resources. One promising development is the creation of "portals" to help users organize the bewildering collection of information resources (Zemon 2001; Mischo 2001). A portal is a Web interface that allows users to customize their personal access to information resources. Examples are *MyLibrary* at North Carolina State University (http://my.lib.ncsu.edu) (Pace 2001); *MyGateway* at University of Washington Libraries (https://www.lib.washington.edu/Resource/login.asp?user=guest); and *Zportal* software designed by the Fretwell-Downing Company (http://www.fdgroup.com/fdi/zportal/about.html). Portals enable searching across multiple information repositories, and associated user profiles enable customized current awareness updates. Challenges include protecting user privacy while maintaining user preference profiles, finding money and time to develop the complex database systems, and integrating with the commercial sector to enhance functionality of the systems (Machovec 2001).

Another challenge for all disciplines is dealing with the cost of electronic journals. Many patrons assume that electronic journals cost less than print. The whole cost issue is rather complex. Some e-journals such as AGU's *Geochemistry, Geophysics, Geosystems* ($G^3$) start out free, but then move to paid subscriptions. Some e-journals originally come free with a paid print subscription, but later are no longer free. Patrons become used to the online access and are frustrated when they are no longer available. The additional charges for the online versions compound the already escalating costs of journals. Publishers assume that libraries will be able to absorb the increased costs for the online version by canceling duplicate print subscriptions when, in reality, many libraries have already cancelled duplicate subscriptions. Libraries are still reluctant to cancel the last print copy due to concerns about archiving.

Publishers fear that revenues will be lost when individual subscribers cancel subscriptions due to the availability of e-journals at universities, and the publishers expect the libraries to make up the differences.

In some cases, costs can be lowered by signing long-term agreements, and by forming consortia, but there are added challenges of inter-institutional decisions of who will cancel which title, and what titles will be added. What happens when an institution leaves the consortia? What happens if the publisher raises the prices beyond what the institutions can afford? Signing long-term agreements with publishers may result in reduced options for decisions and forced cancellation of other valuable journals when cuts are necessary.

There has been some experimenting with in-house dissemination of intellectual output as a way to lower costs and take back ownership of intellectual property. MIT Libraries and the Hewlett Packard Company have joined to produce Dspace, a "digital system that captures, preserves and communicates the intellectual output of MIT's faculty and researchers" via the Web (Stuve 2001). Another answer to spiraling journal costs is SPARC, the Scholarly Publishing and Academic Resources Coalition (http://www.arl.org/sparc/). SPARC is a worldwide coalition of research institutions and library consortia supporting increased competition in scholarly publishing by encouraging and supporting new low cost scholarly journals (Buckholtz 1999). SPARC partners related to geoscience include Columbia Earthscape (http://www.earthscape.org/) and Geochemical Transactions (http://www.rsc.org/is/journals/current/geochem/geopub.htm).

Crawford has examined the current state of free, electronic refereed journals (Crawford 2002). Of the 104 titles listed in the *1995 Directory of Electronic Journals, Newsletters and Academic Discussion Lists*, more than half were still publishing in 2001. Twenty-six of the survivors are affiliated with or sponsored by universities or associations. Few are in the areas of science and medicine. According to Crawford, "refereed scholarly journals are hard to maintain without any revenue, and it has been difficult to use electronic publications for tenure or to show their impact on a field."

The previously mentioned problem of archiving e-journals has not been fully resolved. One encouraging example is JSTOR (Journal Storage), a reasonably priced, quality, electronic archive of scholarly journals. Although JSTOR currently has very few titles related to geology, they are surveying librarians to determine future additions to the collection. Another example is the recent cooperative initiative between the American Physical Society (APS) and the Library of Congress (Ameri-

can Physical Society 2001). They have created the Physical Review Online Archive (PROLA) (http://prola.aps.org/). Launched in 1998, it has
been expanded to include APS journals back to 1893. Astrophysics
Data System (ADS) (http://adswww.harvard.edu/), funded by NASA
and hosted by the Harvard-Smithsonian Center for Astrophysics, provides free access to the full text of articles in astronomy and astrophysics.
Hopefully many other organizations, including geological societies,
will follow these examples and provide online access to early archives
at a reasonable price. This would be a most welcome trend.

Another potential problem with electronic journals is related to citation of articles. As publishers move toward the electronic format, the
electronic version is sometimes becoming the "version of record."
American Geophysical Union (AGU) is currently using "digital object
identifiers" (DOIs) for citing and retrieving articles instead of using
page numbers (Mader 2002). Some fear that this practice may create
confusion for those who do not have access to the electronic version, in
addition to awkward and error-prone citations, and related problems for
interlibrary loan departments. Others point out that the DOI can have a
logical form similar to current citations. For example, a DOI could consist of a publisher number identifier followed by an abbreviated journal
title, then volume number, issue number, and finally article number.
However, the DOI is inherently a "dumb" identifier and publishers are
assigning DOIs that do not include journal name, volume, or page number. The DOI is a relatively new phenomenon, and it may take a while
for standards to emerge and for users to adjust.

A number of questions should be addressed when making decisions
concerning format of journals. Should the library purchase print only,
online and print, or online only? Is the online version included at no ex-
tra charge with the print version? If not, can the print version be cancelled to keep the cost the same? If the print version is cancelled, will
there be perpetual access to the information? Can libraries count on the
publisher to make it available? Can the campus legal counsel accept the
license agreement? Will there be unrestricted access by IP address? Are
all parts of the publication included in the online version? How well
does the online version function? How clean are the tables, figures, and
illustrations? Is it part of a consortia agreement? If so, what are the ram-
ifications of that? Which e-journals should be cataloged? Some of these
questions may be resolved as librarians voice concerns and opinions to
providers. As time passes, some standardization can be expected, which
will also reduce the number of questions. The answers to other ques-

tions will depend on the budgets and needs of individual libraries and institutions.

## E-Books

In addition to e-journals, the option of selecting books in electronic format is also emerging. Books that lend themselves favorably to electronic format are those that are often used by a large number of people, but for which only small parts are needed, for example, reference material. Computer science researchers use e-books to quickly answer questions about programming. Online encyclopedias such as *Britannica* are very useful for most disciplines. On the other hand, some resources such as an online Gazetteer might not get enough use at a smaller institution to justify the cost. While online subscriptions might not be cost effective for most geological reference material, the one-time cost of CD-ROMs might be acceptable for works such as *Glossary of Geology* (Jackson, 1997), *Manual of Mineralogy* (Dana et al., 1997), or any of the volumes of the *Encyclopedia of Earth Science Series* (Chapman & Hall/Kluwer). There are library issues related to the use of reference material in CD-ROM format including whether or not the material will be circulated, whether the CD-ROMs can be networked, the availability of stand-alone equipment, and user issues connected with the differences in the search interfaces and retrieval engines. The problem of cataloging will be covered later in this paper.

## Geographical Information Systems (GIS)

Geographic information is another area of electronic choice, and Geographic Information Systems (GIS) is an enduring topic of interest. Increasingly, geographical information is being produced in digital format, and this information can be layered through the use of GIS. New relationships become apparent, and new products can be created. For example, geoarchaeologists might be interested in combining data related to soil types, locations of river features over time, and known archaeological sites to create a new map that would help detect undiscovered sites. The librarian is faced with a number of considerations. Should the library purchase/collect these materials? Should the library also purchase and maintain the hardware and software necessary to use the data? How much technical support should the staff provide to users? Should the CDs be checked out? Can data be stored on servers? Can it be made Web accessible? When cataloging the material, what information, such as type of map projection, should be included? Deckelbaum has addressed a number of these concerns (Deckelbaum 1999). According to Derksen et al., answers to these and other questions must be specific to institutions, and driven by local need (Derksen et al. 2000). For example, the University of Michigan's map library has begun to load more frequently requested GIS data on servers, allowing remote access to users with a valid university IP address. Other data can be requested using a Web form. Onsite help is available to novice users and those with complex problems and requests (Jensen 2000).

### Current Awareness

#### Journal Articles

While the technology revolution presents us with the challenge of choices, the continued exponential increase in knowledge presents the challenge of awareness: What information is available and how do researchers find it? The increase in published material combined with serials cuts make current awareness a serious problem for most researchers. One answer is to make available multidisciplinary online index/abstracting tools such as *Current Contents.* Another is to provide access to tables of contents services such as *Reveal,* which e-mails tables of contents of selected journals, or e-mails the titles of relevant articles by subject. There is a trend for publishers to make tables of contents of their journals available free on the Web. Librarians often know the areas of interest of researchers, and forward any relevant titles they may come upon. The next step, of course, is to acquire the article. As budget cuts force journal cancellations, this step becomes increasingly challenging. As more journals become available in electronic full-text format, current awareness services can provide researchers with both the citation and a link to the full-text of the article (Schlembach 2002). Libraries generally provide interlibrary borrowing and document delivery services to provide access to journal articles not owned by the library. Researchers and scholars are aware of others working in their subject areas, and often share information.

#### Data Sets

Finding actual research data collections can be very difficult. However, specialized finding tools are being created for data. Leicester and Major have described NASA's *Global Change Master*

*Directory* (http://gcmd.nasa.gov), a free, searchable directory of data collections related to a wide range of earth science topics (Leicester and Major 2000). Researchers are encouraged to submit metadata records of their data collections, and collaborative efforts are under way with various societies, agencies, and publishers.

A related concern is the difficulty in finding what geospacial data collections exist, and how to acquire them (Bier 2000). Currently, GeoRef, OCLC, and library catalogs are very incomplete in regard to geospacial data sets. The *National Geological Mapping Database* is still being built. Currently, finding aids include specialized Web directories, publishers' Web sites, and calling or e-mailing state or Federal agencies that would likely produce or house the data. Badurek has reviewed the current field of geographic information science and information sharing, and has discussed possible future developments (Badurek 2001). The Digital Earth Project, headed by NASA and involving more than 30 U.S. federal agencies, was initiated in 1998 (Robinson 2000). The project was envisioned to collect, integrate, and display in 3-D all the geological, geographical and demographic data of the Earth.

## Cataloging

A problem related to awareness is the current state of cataloging electronic resources. Adding links for e-journals and e-texts, such as those offered by the National Academy Press, to the Online Public Access Catalog (OPAC) records can be very time consuming. Medeiros has discussed library cataloging in the "Internet age," including the creation of individual library portals (Medeiros 1999). Providing access to the myriad Internet resources is even more daunting. Ward and VanderPol have discussed the current problems and models, and also have posed a temporary solution (Ward and VanderPol 2000). Currently neither the available search engines nor the OPAC are adequate for finding Internet resources. Problems with adding Internet resources to the online catalog include the instability of Internet sites, the paucity of copy cataloging (which necessitates original cataloging), and the time and training necessary to find and evaluate Internet sites. Current models for dealing with the problems include OCLC's Cooperative Online Resource Cataloging (CORC) project in which participating institutions contribute records of Internet sites to a shared database, and also create metadata. Other initiatives include the Dublin Core, which provides a limited number of standard fields designed specifically for Internet resources, and PURL (Persistent Uniform Resource Locators) which counter the problem of migrating Web sites by pointing to the current URL of a specific Internet resource.

Ward and VanderPol have proposed a staged approach to standardized access to Internet resources (Ward and VanderPol 2000). They suggest that, initially, the subject lists created by library staff be treated as collections. The Internet resources included in the subject lists have already been evaluated and selected by librarians, and links could easily be checked regularly. Ward and VanderPol suggest, as an intermediate solution, that libraries catalog their Web pages that provide lists of Internet resources.

## Preservation and Dissemination of Information

Another persistent challenge is preservation of material. While geoscientists depend on current information, the older material is still very important to them. International material is also very important in the geosciences. Wishard and Musser have compared old and new preservation techniques in regard to geoscience materials and discussed the need to prioritize materials that are candidates for preservation (Wishard and Musser 1999). The electronic format can be a good preservation tool. Many old geology monographs, such as the United States Geological Survey (USGS) publications, are in danger due to aging of the material, theft, and mutilation. An answer may be to digitize the publications and make them available on the Web. This would be an important public service. On one hand, the presence of these documents on the Web could draw attention to the original documents and instigate more use and theft. On the other hand, perhaps the original could be withdrawn from circulation and stored in a protected area if the information is available on the Web.

Efforts are being made to digitize older maps for preservation, and this should continue. To effectively view digitized maps, powerful workstations and sufficient computer memory is required. Large color printers are very expensive, and, unless used often, ink dries up and creates expensive, time-consuming maintenance problems. In addition, basic bibliographic control is often lacking for digitized maps. More work needs to be done in these areas.

There are certain general problems to consider when using electronic format for preservation. As technology continues to change, new formats will emerge. Libraries should maintain a program to detect resources that are becoming unusable due to changing technology. It may be necessary to refresh or migrate data to new technologies, or to keep older software and hardware. For example, North Dakota State University library staff members have recently found 5-1/4″ floppy disks included with monographs. Not a single computer has been found in the library that can read and display the information on these disks. Costs for migrating or refreshing data include funding for the process, but also staff time spent in deciding which data should be converted. Decisions could depend upon input from subject librarians, use statistics, and input from researchers in the subject area.

## Reference and Information Literacy

### Reference Tools

The driving forces also present challenges and opportunities in the areas of reference and information literacy. The digital format lends itself very well to the modification and creation of reference tools such as online, searchable indexes. One example is the *Collective Bibliography of North Dakota Geology* (http://dp3.lib.ndsu.nodak.edu/ndgs/), a combination of three print bibliographies and subsequent updates. The print resources were converted to digital format, and the data were imported into a relational database called MySQL. The database is searchable using a microcomputer running UNIX. Common Graphic Interface (CGI) forms were created in-house using the Perl programming language. These forms offer flexible interface designs for database users and programmers (England et al. 2000). Since the software was created in-house, the only costs were employee time and computer memory. Other examples of locally created searchable reference tools are Kansas Geological Survey's *KGS Online Bibliography of Geology* (http://magellan.kgs.ukans.edu/Bib/index.html), and Geoscience Information Society's *Union List of Field Trip Guidebooks of North America Online* (http://www.georef.org/gnaintro.html).

DeFelice has described The Digital Library for Earth System Education, funded in part by the National Science Foundation (NSF) (DeFelice 2000). This will be a specialized digital library designed to increase dissemination and use of earth science educational materials. Tahirkheli and Andrews have described the conversion of *The Arctic Bibliography* from a print to a digital product (Tahirkheli and Andrews 1999). Myers has described a Web application developed at the University of Wyoming, which manages, stores, retrieves, and facilitates use of multimedia geosciences resources (Myers 2000). Badurek described a system for information retrieval from very large databases, termed "information visualization" (Badurek 2000). It is essentially a graphical user interface that represents data holdings as a "landscape" or "information space." The time spent envisioning, creating, maintaining, and discovering online reference tools will continue to increase for geology and other science librarians.

## Instruction

The digital revolution is also impacting library instruction. The myriad resources and formats can be bewildering, especially to freshman students. Fleming points out that bibliographic instruction used to consist of a tour of the library and instruction in the use of the card catalog and the major print indexes (Fleming 2001). Today, library instruction often occurs in a high-tech classroom and involves teaching computer skills, electronic search techniques, and evaluation of Web resources. Yocum and Almy suggest that, with these new resources, more instruction is needed rather than less (Yocum and Almy 1999). Problems occur when the proper technology is not available to teach online resources, especially on short notice. Classrooms, especially computer labs, may already be booked, online resources may have a limited number of simultaneous users, networks may become overloaded, and the system may be "down" at the critical moment that it is needed. It is wise to have several backup plans, including a very low-tech version such as overheads.

It is useful to offer library instruction geared to particular class assignments. The Web is an effective tool for teaching information literacy. Most of the information resources needed for a particular assignment can be gathered together on one Web page, with links to electronic resources. The Web page can be used during instruction sessions, and then can be accessed at any time by the students. The Web page can be linked to the instructor's Web page for the particular class. Much of the information may be re-used for pages related to other classes, with modifications including examples targeting a particular assignment. An example for a petrology and petrography class assignment at the University of Illinois at Urbana-Champaign can be found at <http://www.library.uiuc.edu/gex/Classes/Geol336.html>.

## USGS and Other Government Agencies

Another challenge facing geology libraries is the tenuous nature of the United States Geological Survey (USGS) and other government agencies. The USGS is a major supplier of free or low cost, high quality geological information. In recent years the USGS and other government agencies have been targets of budget cuts. Cutting or dissolving the USGS would have a major impact on geology libraries. Hopefully all librarians and users of USGS resources will voice their concerns about this potential loss. Other related concerns include the status of the Federal Library Depository Program (FLDP), and the complexities associated with managing electronic publications of the government agencies including indexing, storage and preservation in the FLDP.

The vulnerability of government publications has been highlighted by several recent events. After the September 11th attack on the World Trade Center, at least one USGS document (a CD-ROM) was recalled due to national security (Blair 2002). All federal agencies were asked to remove sensitive information from the Web (Sammon 2002). In another unrelated incident, all of the Department of Interior Web sites were closed down on December 5, 2001 due to a lawsuit (Griles 2001). The USGS is a part of the Department of Interior. Due to the critical nature of USGS information, including earthquake and tsunami warning systems, access to many of the USGS Web sites was reinstated on December 8, 2001 (Lamberth 2001). As of April 9, 2002, access to many of the other Department of the Interior Web sites was still restricted, although some access was enabled through private Internet providers.

Several recent studies have detected problems related to accessing government documents in an online environment. Derksen found a lack of comprehensive coverage of both electronic and print USGS publications by traditional indexing sources (Derksen 2001). Another case study was conducted by Jensen (Jensen 2002). After diligently searching, Jensen discovered a number of online documents that were not consistently indexed by any of the major indexing tools. Although the USGS is working to resolve this problem, the issue of hidden government documents and Web resources is likely to remain for some time (Blair 2002). Klaus has raised additional questions related to the migration of government documents from print to cross-media format: "What is the 'archive' version of the publication once archive-quality books are not produced? Is a manuscript prepared in different mediums the same publication?" (Klaus 2000).

## *Legislation and Copyright*

Another challenge for all librarians is awareness of current legislation and copyright law. The shift to digital format has been accompanied by changes in copyright legislation such as the Digital Millennium Copyright Act of 1998 (Davis and Fiander 2001). There are several recent important developments in the areas of United States copyright and legislation. Several recent bills in the U.S. House of Representatives have been based on concern about electronic database piracy, but they could also curtail access to geophysical information (Folger 2000). Another recent critical topic is distance learning legislation (Pike 2001). The Technology, Education and Copyright Harmonization Act of 2001 was introduced to address distance learning limitations included in the Copyright Act of 1976.

Johnson examined 76 geology journals to determine changes in the nature of copyright and permission statements between the years 1990 and 2000 (Johnson 2000; Johnson 2002). The study indicated that approximately 10% more geology journals included copyright permission statements in 2000 than in 1990. In addition, there was an increase in mention of copyright law during the same time period. Copyright permissions were slightly more favorable to authors in 2000 than in 1990. Johnson suggests that the differences may be partially related to initiatives such as the previously mentioned Scholarly Publishing and Academic Resources Coalition (SPARC) that work toward strengthening the proprietary rights of authors.

Geoscience librarians should be aware of current developments in the areas of legislation and copyright, and work to maintain the widest possible access to geologic information. Copyright laws differ by nations, such as Canada and countries within Europe, and it is important to be aware of the interactions and influences of international law (Harris 2001; Battisti 2001).

## Digital Librarianship and User Services

Many subject librarians are spending an increasing amount of their time initiating and supervising digital projects. Digital projects have the potential to benefit library patrons, the libraries, and librarians. However, it is also possible to spend time and effort creating a product that will not be used. Care should be taken to determine real patron information needs that could best be served by a digital product. Digital products can also benefit the librarian. Digital products can be time consuming both to create and to maintain. However, a lot of time is saved if the product will be used by many patrons, for a number of classes, and over a number of years with some revision. For example, a particular course may have an assignment requiring a paper on a particular topic, such as Quaternary dating methods. The time required to create and update a Web page of resources is more than compensated by the time needed to help numbers of patrons look up the same resources each semester or year. For another example, the Collective Bibliography of North Dakota Geology (http://dp3.lib.ndsu.nodak.edu/ndgs/) was initiated because GeoRef does not index all the material of interest to North Dakota geologists, and indexing of some sources has a long time lag.

The creation of digital products will not replace personal contact. Indeed, more contact may be necessary. Personal interaction with faculty, classes, and students will help determine what products will be useful, and will ensure a higher quality product. Digital projects can be expected to benefit both on-campus users, as well as distance users. When creating online products, both types of users should be considered.

In addition to reference and information literacy projects, digital projects may include preservation of knowledge and wider dissemination of information. Multiple projects should be prioritized. Workflow will be partly determined by available budget and staff.

Digital projects and user services ventures often require teamwork. The librarian may work in conjunction with various computer science experts, depending on the project. Collaborators may include computer science students or a permanent library employee with computer science expertise. For projects requiring the creation of a database and search engine, a computer expert may create the search engine, student workers or clerks may digitize the information and populate the database, and the librarian will edit the database and consult with the computer expert about the final product. Digital projects have a life of their own; once created they require ongoing attention. The more Web pages and Web products that have been created, the more time will be necessary to maintain and update them. Web counters can help to gauge use and determine whether a site should be maintained or advertised in some manner to increase use.

## CONCLUSIONS

There is a quotation, similar versions of which are attributed both to Presidents Eisenhower and Ford: "Things are more like they are now than they [ever] have been [before]." No matter how fast things change, or in what direction, at least that much will always remain true.

Some are concerned that, with the digital revolution, e-journals, e-books, and e-indexes, the role of libraries and librarians will be reduced. While that may be true eventually, it will most likely not be happening any time soon. There is an old science fiction story about a future time when scientists had become so specialized that a new profession was created called "generalist." The task of the generalist was to know a little about many different subjects, and help bridge the gap between specialists by finding related information. The library professional will increasingly fill that role. Geology librarians will continue as information specialists, helping to prepare users for a lifetime of learning, working to standardize the various electronic resources, and ensuring that knowledge is preserved in some format.

## REFERENCES

American Physical Society. 2001. About PROLA. http://prola.aps.org/about.html (accessed 3/24/01).

Badurek, C.A. 2000. Information visualization approaches for the geosciences. *Proceedings of the 35th Meeting of the Geoscience Information Society, November 11-15, 2000, Reno, Nevada*, p. 23-27.

Badurek, C.A. 2001. Geographic information science and technologies: Impacts on information access and exchange for the geosciences. *Geological Society of America Abstracts with Programs*. 33(6):A61.

Battisti, M. 2001. The future of copyright management: European perspectives. *IFLA Journal*. 27(2):82-86.

Bier, R.A., Jr. 2000. Where are the maps? Or the changing ways to find maps in the age of GIS and the Web. *Proceedings of the 35th Meeting of the Geoscience Information Society, November 11-15, 2000, Reno, Nevada*, p. 55-59.

Blair, N. 2002. News from the USGS libraries. *GIS Newsletter*. 195:11.

Buckholtz, A. 1999. SPARC: The scholarly publishing and academic resources coalition. *Issues in Science and Technology Librarianship*. Spring 1999: article 2. http://www.library.ucsb.edu/istl/99-spring/article2.html (accessed 4/28/02).

Bush, V. 1945. As we may think. *Atlantic Monthly*. 176 (1):101-108. (Reprinted in *Library Computing*. 18(3):180-188).

Chan, W. 2001. Creative applications of a Web-based e-resource registry. *Science & Technology Libraries*. 20(2/3):45-56.

Crawford, W. 2002. Free electronic refereed journals: Getting past the arc of enthusiasm. *Learned Publishing*. 15(2):117-123.

Dana, J.D., Dana, E.S., and Gaines, R.V. 1997. *Dana's New Mineralogy: The System of Mineralogy of James Dwight Dana and Edward Salisbury Dana*. 8th ed. New York: Wiley.

Davis, T.L and Fiander, P.M. 2001. The Digital Millennium Copyright Act: Key issues for serialists. *Serials Librarian*. 40(1/2):85-103.

Deckelbaum, D. 1999. GIS in libraries: An overview of concepts and concerns. *Issues in Science and Technology Librarianship* (winter 1999). http://www.library.ucsb. edu/istl/99-winter/article3.html (accessed 8/16/00).

DeFelice, B.J. 2000. Building a community centered digital library for earth system education. *Proceedings of the 35th Meeting of the Geoscience Information Society, November 11-15, 2000, Reno, Nevada*, p. 91-95.

Derksen, C.R.M., Sweetkind, J.K., and Williams, M.J. 2000. The place of geographic information system services in a geoscience information center. *Proceedings of the 35th Meeting of the Geoscience Information Society, November 11-15, 2000, Reno, Nevada*, p. 29-47.

Derksen, C.R.M. 2001. USGS publications: Current access via the Web and via catalogs. *Geological Society of America Abstracts with Programs*. 33(6):A61.

Duranceau, E.F. 2001. Scaling the Tower of Babel: Challenges in library acquisition of digital resources. *Geological Society of America Abstracts with Programs*. 33(6): A60.

England, M., Joseph, L., and Schlecht, N.W. 2000. A low cost library database solution. *Information Technology and Libraries*. 19(1):46-49.

Fleming, A.C. 2001. Bibliographic instruction for the geoscience undergraduate: A digital wonderland or lost in space? *Geological Society of America Abstracts with Programs*. 33(6):A60.

Folger, P.F. 2000. Database piracy and access to geophysical data; Where is congress headed? *Proceedings of the 35th Meeting of the Geoscience Information Society, November 11-15, 2000, Reno, Nevada*, p. 71.

Griles, J.S. 2001. Memorandum: Disconnection of Internet Service. Office of the Secretary, United States Department of the Interior: http://www.doi.gov/news/grilesmemo. html (accessed 4/9/02).

Harris, L.E. 2001. An excerpt from 'Canadian Copyright Law,' third edition. *Information Outlook*. 5(2):30-34.

Hiller, S. 2002. How different are they? A comparison by academic area of library use, priorities, and information needs at the University of Washington. *Issues in Science and Technology Librarianship* (Winter). http://www.istl.org/istl/02-winter/article1. html (accessed 4/8/02).

Jackson, J.A. 1997. *Glossary of Geology*. 4th ed. Alexandria, VA: American Geological Institute.

Jensen, K.L. 2000. Creating resources for GIS support to remote users. *Proceedings of the 35th Meeting of the Geoscience Information Society, November 11-15, 2000, Reno, Nevada*, p. 49-54.

Jensen, K.L. 2001. Providing access to online government documents in an academic research library collection. *Science & Technology Libraries*. 20(2/3):15-26.

Johnson, K.G. 2000. The nature of copyright permissions in geology journals. *Proceedings of the 35th Meeting of the Geoscience Information Society, November 11-15, 2000, Reno, Nevada*, p. 69.

Johnson, K.G. 2002. Communication with the author, 7 May.

Klaus, A.D. 2000. Migrating ODP 'Proceedings' from print to cross-media publication formats. *Proceedings of the 35th Meeting of the Geoscience Information Society, November 11-15, 2000, Reno, Nevada*, p. 73-90.

Lamberth, R.C. 2001. Order providing partial relief from temporary restraining order. United States District Court for the District of Columbia. http://www.seismo-watch. com/EQS/AB/2001/011207.DOI_Shutdown/011208.PartialRelief.pdf (accessed 4/9/02).

Leicester, S.M., and Major, G.R. 2000. Discovering geoscience data through NASA's Global Change Master Directory. *Proceedings of the 35th Meeting of the Geoscience Information Society, November 11-15, 2000, Reno, Nevada*, p. 63-67.

Licklider, J.C.R. 1965. *Libraries of the future*. Cambridge, MA: M.I.T. Press.

Licklider, J.C.R., and Taylor, R. 1968. The computer as a communication device. *Science & Technology*. 76:21-31.

Machovec, G.S. 2001. Library portals: Customizing and focusing the user's experience. *Libraries and Microcomputers*. 19(1):1-3.

Mader, C. 2001. Current implementation of the DOI in STM publishing. *Science & Technology Libraries*. 21(1/2): 97-118.

Manson, C.J. 2001. The evolution of a state geological survey library–The more things change, the more they stay the same. *Geological Society of America Abstracts with Programs*. 33(6):A60.

Medeiros, N. 1999. Driving with eyes closed: The perils of traditional catalogs and cataloging in the Internet age. *Library Computing*. 18(4):300-305.

Mischo, W.H. 2001. Library portals, simultaneous search, and full-text linking technologies. *Science & Technology Libraries*. 20(2/3):133-148.

Myers, J.D. 2000. Managing local media resources: The Earth Science Media Gallery (ESMG). *Geological Society of America Abstracts with Programs*. 32(7):A-203.

Noga, M.M. 2002. President's Column. *Geoscience Information Society Newsletter*. 195:1, 3.

Pace, A.K. 2001. Should MyLibrary be in your library? *Computers in Libraries*. 21(2):49-51.

Pike, G.H. 2001. A busy year ahead for Congress. *Information Today*. 18(5):16-18.

Regan, C.L. 2000. Digital Earth: The state of geoinformatics, 2000. *Proceedings of the 35th Meeting of the Geoscience Information Society, November 11-15, 2000, Reno, Nevada*, p. 1-22.

Robinson, B. 2000. Creating a digital earth. *Federal Computer Week*. 14(25):17-20.

Sammon, B. 3/21/2002. Web sites told to delete data. *The Washington Times*. http://www.washtimes.com/national/20020321-16859342.htm (accessed 4/9/02).

Schlembach, M.C. 2001. Trends in current awareness services. *Science & Technology Libraries* 20(2/3):121-132.

Stuve, D.H. 2001. Dspace: Meeting the challenge of capturing and preserving MIT's intellectual output. *Geological Society of America Abstracts with Programs*. 33(6):A61.

Tahirkheli, S.N., and Andrews, M. 1999. The Arctic Bibliography: A resource renewed. *Proceedings of the 34th Meeting of the Geoscience Information Society, October 25-28, 1999, Denver, Colorado*, p. 3-9.

Waldrop, M.M. 2001. *The dream machine: J.C.R. Licklider and the revolution that made computing personal*. New York: Viking Press.

Ward, D., and VanderPol, D. 2000. Librarian, catalog thy work! Getting Started integrating Internet resources into OPACs. *Journal of Internet Cataloging*. 3(4):51-61.

Wishard, L.A., and Musser, L.R. 1999. Preservation strategies for geoscience literature: New technologies for an old literature. *Library Resources & Technical Services.* 43(3):131-139.

Yocum, P.B., and Almay, G.S. 1999. Information literacy in the geosciences: Report of a practical inquiry. *Proceedings of the 34th Meeting of the Geoscience Information Society, October 25-28, Denver, Colorado,* p. 15-22.

Zemon, M. 2001. The librarian's role in portal development: Providing unique perspectives and skills. *College & Research Libraries News.* 62(7):710-712.

# Bridge Beyond the Walls:
# Two Outreach Models
# at the University of California, Santa Cruz

Catherine Soehner
Wei Wei

**SUMMARY.** Online access to databases, journal articles, and books has resulted in less frequent contact with our users. As a result, librarians must make greater efforts to reach out to their constituencies to deliver information about new resources, searching tips and techniques, and the state of the library's collections. This article describes and compares the success of two outreach events held by the Science & Engineering Library at the University of California, Santa Cruz (UCSC). Practical considerations such as a time, location, and faculty involvement are discussed. A new marketing tool is introduced which involves person-to-person contact to advertise events. *[Article copies available for a fee from The Haworth Document Delivery Service: 1-800-HAWORTH. E-mail address: <docdelivery@haworthpress.com> Website: <http://www.HaworthPress. com> © 2001 by The Haworth Press, Inc. All rights reserved.]*

**KEYWORDS.** Marketing, faculty outreach, library events

Catherine Soehner, MLS, is Head, Science & Engineering Library (E-mail: soehner@cats.ucsc.edu) and Wei Wei, MLS, MA, is Computer Science Librarian (E-mail: wwei@cats.ucsc.edu), both at the University of California, Santa Cruz, CA 95064.

[Haworth co-indexing entry note]: "Bridge Beyond the Walls: Two Outreach Models at the University of California, Santa Cruz." Soehner, Catherine, and Wei Wei. Co-published simultaneously in *Science & Technology Libraries* (The Haworth Information Press, an imprint of The Haworth Press, Inc.) Vol. 21, No. 1/2, 2001, pp. 87-95; and: *Information Practice in Science and Technology: Evolving Challenges and New Directions* (ed: Mary C. Schlembach) The Haworth Information Press, an imprint of The Haworth Press, Inc., 2001, pp. 87-95. Single or multiple copies of this article are available for a fee from The Haworth Document Delivery Service [1-800-HAWORTH, 9:00 a.m. - 5:00 p.m. (EST). E-mail address: docdelivery@haworth press.com].

10.1300/J122v21n01_08

## INTRODUCTION

Advances in technologies and the Internet have taken library research beyond the walls of the library. The possibility of conducting literature searches away from library in the office or at home is changing the role of the academic library, as has never seen before. "Within the automated architectures of proliferating technology we are sought after for our intellectual methods and communication model" (Herold 2001). Like all information professionals, the science librarians at the Science & Engineering (S&E) Library at UCSC are facing enormous challenges in helping library users meet their research needs and are exploring new ways to promote library services. "In an information age, where information itself is key to performing work tasks, understanding how telecommuting knowledge workers find, use, and depend on information is key to good management" (McInerney 1999). Including faculty, graduate and undergraduate students, the science population at UCSC is approximately 2,157. The S&E Library holds drop-in orientations for its users each fall. In the earlier years, the attendance at these orientations was considerable. However, in the last few years, the attendance has declined and recently dropped to one or two people at each session, due to changing behavior regarding how users access library resources in this networked environment. Clearly, traditional methods to conduct and market library services no longer worked well. Therefore, the S&E Library decided to bring events closer to users and try aggressive marketing techniques to reach out to its research community.

During recent years, the S&E Library has conducted outreach events based on two different models specifically to connect with users around common issues. The first outreach effort was to provide a Science Media Fair outside the library at the campus Sinsheimer Laboratories that introduced faculty and graduate students to recently acquired databases; the second was to hold the 10th Anniversary Seminar, "Scholarly Publishing in Higher Education" in the S&E Library. The purpose of this article is to compare these experimental models and to examine what effect time, location, faculty involvement, promotional strategies, and partnerships had on the number of people who attended each event.

## SCIENCE MEDIA FAIR

"With the advent of online databases, other electronic resources, new methods of document delivery, and access to information, the role of the academic library has begun to change. Students do not have to be physically present in the library in order to access the library's resources. The Internet has opened the resources of library to students and faculty worldwide" (Simmonds 2001).

The first outreach effort was to provide a Science Media Fair outside the S&E Library at the campus Sinsheimer Labs. This event introduced science faculty and graduate students to recently acquired databases. The Science Media Fair took place over a three hour time period during which six half hour presentations were given on various S&E Library and Media Services products and services. The initial planning began in December 1999 with the creation of a Planning Task Force, which consisted of two representatives from the S&E Library and three representatives from Media Services, a unit within the University Library. A careful study was conducted to determine normal traffic patterns during class inter-sessions between the Sinsheimer Labs and other science buildings, taking note of the placement of coffee carts and other obvious gathering places. The task force also made an effort to examine when the science faculty meetings and departmental seminars would be held during the Spring Quarter. The acquisition of this information was critical to maximize attendance and minimize potential conflicts. Based on that knowledge, the task force decided to schedule the Science Media Fair on Wednesday, April 6, 2000 from 11:00 a.m. to 2:00 p.m. in the Sinsheimer Labs conference room and lobby in order to maximize student and faculty attendance.

Partnerships with Media Services, science faculty, and a database vendor added to the success of the event. In addition to participating on the Planning Task Force, the Media Services Department had extensive knowledge regarding Ethernet connections and was expert at setting up multiple computers with access to the Internet in short notice. These skills were especially useful on the day of the event. There were a total of six formal classroom presentations, each lasting 30 minutes. Three presentations focused on S&E Library services and the other three provided information on Media Services applications and products. Two of these sessions, EndNote and Classroom Multimedia Projects, were delivered by faculty members. The former was presented by a professor of chemistry who utilized EndNote databases for his research; the latter was outlined by a faculty member who conducted multimedia projects with the assistance of Media Services staff. Media Services also offered Classroom Media Information and Distance Education classes and the science librarians provided demonstrations on SciFinder Scholar and Web of Science.

While the six formal presentations were held in a conference room, the S&E Library and the Media Services staff also provided informal, online demonstrations of science databases and Media Services products in the Sinsheimer Labs lobby area, which is located very close to the conference room. Several workstations were set up and refreshments were generously provided by the Chemical Abstracts Service, the makers of SciFinder Scholar. As a result of this effort, 150 people participated in the event, many of whom were unable to find a seat during the Web of Science, EndNote, and SciFinder sessions. In the months immediately following the Science Media Fair, we noticed an increase in the use of EndNote which was demonstrated by an increase in the number of questions at the reference desk and an increase in the attendance of EndNote classes. Additionally, chemistry graduate students and faculty increased their use of SciFinder Scholar as evidenced by questions regarding the installation and use of the software. It was a rewarding experience for the attendees, and a truly successful experience from the library's perspective.

## 10TH ANNIVERSARY SEMINAR

May 2001 marked the 10th year of library services rendered from the "new" S&E Library building. Using the timing of this event as a cornerstone of our outreach effort, the Seminar Planning Group began to organize a faculty seminar to be held within the S&E Library. Since the increasing volume and escalating costs of scholarly communication are concerns of both faculty and librarians, the Seminar Planning Group, which consisted of four science librarians, chose Scholarly Publishing in Higher Education as the seminar theme. Knowing that many of the library's journals were accessible online and no longer required a user to enter the library, the seminar provided an opportunity to focus attention on the intellectual content of the library rather than the building itself. "The environment for scholarly information is expected to be highly fluid for at least the next decade, as universities attempt to meet the challenges of scholarly and scientific communication in the 21st century" (Vasanthi 2001).

Planning began in February 2001, and in the early stages of the planning, the group engaged in many discussions concerning the topics to be presented and the major issues related to scholarly publishing in higher education. Due to the complicated planning process, an outline was established to aid communication among the Planning Group. Finally, the

Planning Group decided upon and received confirmation from two outstanding speakers, known to be knowledgeable about the subject. One of the speakers, Paul Ginsparg, is well known among physics researchers and was thus a potential draw for science faculty and graduate students. The second speaker, Catherine Candee, is the Director of eScholarship at the California Digital Library and while she was well known to the University of California librarians, her program and its focus were important concepts that the Planning Group wanted the faculty to become more familiar with as they considered the impact of the scholarly communication crisis.

It was clear that partnerships would again add to the success of the overall seminar planning and implementation. Five campus offices were approached for their support and all agreed to be sponsors of the event. The Division of Natural Sciences, the School of Engineering and the Office of Research contributed funds for the speakers. The University Library and the Librarians Association of the University of California, Santa Cruz Division provided refreshments and equipment needed for the success of the program.

The seminar took place from 4:00 p.m. to 6:00 p.m. on a Wednesday in the later spring when it would not conflict with other science faculty events. Since it was to celebrate the 10th anniversary of the S&E Library, the seminar was held in the library's Current Periodicals Room. Targeted to the science faculty, graduate students and librarians, the speakers delivered their presentations on the topics of New Methods of Scholarly Communication and Scholarly Communication and E-print Archive. The topics were right on target and brought many science faculty members together in the S&E Library for the first time, at one single event. The success of this seminar was a truly collective effort with all S&E Library staff helping out in some way during the day of the event. The results of these actions were to be stimulating and encouraging.

## COMPARISONS

If attendance is considered to be the measure of success, then it is clear that the Science Media Fair was more successful than the 10th Anniversary Seminar. While there were many differences between the events, two factors seemed to have influenced the overwhelming attendance at the Science Media Fair: location and timing.

The Science Media Fair was held in the Sinsheimer Laboratories in a conference room and lobby area where a coffee cart is stationed and which was, therefore, a typical meeting place for students and faculty during lunch. Holding the Fair near offices and labs aided attendance since it required less effort for our target audience to find their way to the event. Additionally, Sinsheimer Labs is an area with which many of our faculty and students are intimately familiar and, as a result, it provides a level of comfort that they may not have experienced with the library, especially if they perform most or all of their library research away from the library building.

Timing was another critical factor that contributed to the increased attendance at the Science Media Fair. As mentioned earlier, the Science Media Fair was held near a coffee cart, a space where faculty and students find food, coffee, and a place to eat their lunch. Since the Fair ran from 11:00 a.m. to 2:00 p.m., the event capitalized on the usual traffic flow of the area.

Timing seemed to work against the 10th Anniversary Seminar. First, the date and time were selected to accommodate the speakers rather than to capitalize on a known open time in faculty or student scheduling. Second, there was an unexpected loss of electricity two hours prior to the event during a year of scheduled blackouts in California. While the lack of electricity lasted only an hour, it was sufficient to bring into question whether or not the event would continue as planned. Additionally, such a disturbance in a workday could easily change a person's mind about whether they would attend an event later in the day.

Considering the factors of location and timing, it would seem that the 10th Anniversary Seminar was doomed to failure. However, the attendance of approximately 60 people at the Seminar and 150 people at the Science Media Fair leads one to believe that there were similar factors contributing to this common success: partnerships with other library and campus offices, inclusion of faculty in the program, and an assertive promotional strategy.

As described earlier, the planning groups for the Science Media Fair and the 10th Anniversary Seminar embraced and established partnerships at the early planning stages. For the Science Media Fair, the Media Services staff partnered with the science librarians on the planning task force and through their functional tasks. The success of the 10th Anniversary Seminar was also achieved by partnering with key campus offices such as the Division of Natural Sciences, the School of Engineering, and the Office of Research. Through their financial and promotional support, the S&E Library was able to bring two outstanding speakers, librarians, faculty, and students together to discuss the important issue of the economics of publishing research results.

The inclusion of two faculty members from the UCSC campus in the Science Media Fair and the inclusion of a well-known physics professor in the 10th Anniversary Seminar added to the attractiveness of both of these events. It gave the impression that each of these events was not just focused on library related matters, but a program with relevant information impacting the everyday teaching, learning, and research activities of the University and a program supported, attended, and produced by respected faculty. While it might be easy for librarians to despair about the idea that a faculty member is necessary to make our information attractive in an academic setting, we must begin to realize and embrace that libraries are a service industry. Many service industries use similar advertising techniques to hook their customers. If a marketing technique works, the technique doesn't make our information less legitimate, it just makes us intelligent marketers, using strategies that work to continue success as defined earlier: increased attendance.

## PROMOTION STRATEGIES

A marketing plan was developed for both models. Both plans were similar in that they consisted of several publicity components including design of promotional material such as posters, pencils given away on the day of the event with the S&E Library name and URL, flyers distributed to the science and engineering departments, and Web-based promotional announcements to campus publications. Another promotional method was to e-mail announcements using active e-mail lists for science faculty, graduate students and the entire campus. Lastly, it was also important to note that using refreshments as a promotional tool definitely helped to attract a great number of people to these events.

All of these efforts involved typical marketing techniques and while they contributed to our success, there was another marketing tool used that we believe tipped the scales to increase attendance beyond our expectations for both events. One of the most assertive methods used to market the events was to have a librarian and a staff member hand deliver flyers door-to-door and office-by-office right before and during the events. The library would not have had such a large turnout for the Science Media Fair were it not that a science librarian walked around prior to each half hour session to faculty and graduate students' offices and labs encouraging them to attend. Similar efforts were made when promoting the S&E Library 10th Anniversary Seminar that was held in the Current Periodicals Room of the library. On the day of the seminar, two science librarians and a library staff member spent an hour walking through science offices and labs recruiting attendees. When the electricity had gone out two hours before the event, a librarian walked around again to make sure that faculty, staff and students knew that the event was still going to happen and to re-invite them to the Library for the event. These Herculean efforts increased the number of attendees in a way that more passive marketing tools had not in the past, such as our experience with drop in library orientations.

## *CONCLUSION*

"An immense amount of extremely valuable information is in existence if only one knows where to find it and how to deliver it" (Herold 2001). It is clear that in this digital age, the personal touch is appreciated and garners attention. While it takes more time and a very outgoing personality, the results are extremely gratifying. People want and expect more personalized service and seem to respond positively when they get it.

The theme of personalized services is repeated again in the variables of timing and location. When the time and location were convenient to the students and faculty, they were more willing and able to attend. Requiring the users to come even the short distance between the science buildings and the library was enough to deter many of them. An assertive library marketing plan and event invitations delivered door to door by the library staff was definitely a true asset to the success of these two outreach models.

The impact of telecommunication and changing nature of work has affected library's delivery methods. In order to achieve the outcomes required in this changing environment, the S&E Library has been searching for new ways to reach out to its faculty and student population. This means that the library has to create a range of teaching and learning strategies and provide technologies beyond the library walls, in a variety of locations such as in the offices and the labs, or at dorms and home, at a time most convenient to its user. The barriers and challenges to the successful implementation of informational services can be overcome by forging cooperative partnerships, employing cooperative planning, focusing on strategic timing and last but not least, adding a personal touch.

# REFERENCES

Herold, Ken R. 2001. Librarianship and the Philosophy of Information. *Library Philosophy and Practice* 3(2): 2.

McInerney, Claire R. 1999. Working in the Virtual Office: Providing Information and Knowledge to Remote Workers. *Library & Information Science Research* 21(1): 1.

Simmonds, Patience L. 2001. Usage of Academic Libraries: The Role of Service Quality, Resources, and User Characteristics. *Library Trends* 49(4): 626-627.

Vasanthi, Christina M. 2001. Changing Environment of Academic Libraries: End-User Education and Planning Strategies for Libraries in India. *Library Philosophy and Practice* 4(1): 4.

# Current Implementation
# of the DOI in STM Publishing

## Cynthia L. Mader

**SUMMARY.** The continuing growth of the Internet has revealed the many advantages and strengths that current technologies can offer. However, it is evident that several key issues pertaining to the persistent identification of digital objects have yet to be addressed and new techniques are needed if the Internet is to provide any degree of reliability for robust management of and access to intellectual properties. In recent years several techniques for persistently identifying digital objects and determining consistency within a dynamic digital environment have been proposed. The Digital Object Identifier (DOI) has the potential to provide needed solutions to these issues. This paper will explore the fundamentals of the DOI and how scientific, technical and medical (STM) publishers are currently using it. *[Article copies available for a fee from The Haworth Document Delivery Service: 1-800-HAWORTH. E-mail address: <docdelivery@ haworthpress.com> Website: <http://www.HaworthPress.com> © 2001 by The Haworth Press, Inc. All rights reserved.]*

**KEYWORDS.** DOI, digital object identifiers, electronic publishing, reference linking, CrossRef, OpenURL, STM publishers, science, technology, medical

---

Cynthia L. Mader, MLS, is Physical Sciences Librarian, Brill Science Library, Miami University, Oxford, OH (E-mail: maderc@muohio.edu).

[Haworth co-indexing entry note]: "Current Implementation of the DOI in STM Publishing." Mader, Cynthia L. Co-published simultaneously in *Science & Technology Libraries* (The Haworth Information Press, an imprint of The Haworth Press, Inc.) Vol. 21, No. 1/2, 2001, pp. 97-118; and: *Information Practice in Science and Technology: Evolving Challenges and New Directions* (ed: Mary C. Schlembach) The Haworth Information Press, an imprint of The Haworth Press, Inc., 2001, pp. 97-118. Single or multiple copies of this article are available for a fee from The Haworth Document Delivery Service [1-800-HAWORTH, 9:00 a.m. - 5:00 p.m. (EST). E-mail address: docdelivery@haworthpress.com].

10.1300/J122v21n01_09

## *INTRODUCTION*

The rapid advancement of technology in the past ten years has brought about many new issues and challenges for publishers, librarians, scholars, and other information specialists. The Internet now offers a tremendous amount of flexibility and accessibility that the print world never could provide. However, the electronic environment has not developed without concerns being raised from a multitude of researchers and professionals in diverse fields. In terms of electronic journals, the main issues pertain to electronic rights management, archiving, authenticity, and data consistency. Perhaps the most serious concern stems from a need to achieve the same level of consistency and reliability in document identification originally provided in the print world. One approach currently in development that proposes to address these issues is the Digital Object Identifier (DOI) System. This paper will discuss the fundamentals of the DOI and will address the following questions: What is the DOI? How does it work? What applications does it have? How are the scientific, technical and medical (STM) publishers currently utilizing the DOI?

## *BACKGROUND*

In 1996, the American Association of Publishers (AAP) formed a committee to address the issues and concerns of intellectual property management as well as facilitate access to and use of these intellectual properties in the digital environment. It quickly became apparent that tackling the intellectual property rights issue as a whole was too large a task at that time. Instead, the AAP decided to focus their efforts first on the smaller, yet very critical task of creating a persistent identifier to assist in the identification of objects within the ever-changing environment of the Internet (Rosenblatt 1997; Davidson and Douglas 1998). It is widely recognized that the Uniform Resource Locator (URL) has not proven to be a consistent identification entity (Lynch 1997; Caplan and Arms 1999; Caplan 2001a; Davidson and Douglas 1998). The reorganization of Web sites often results in the frequent, yet frustrating "Error 404–File Not Found" message that many Internet users are so accustomed to seeing today. It seems obvious that the development of an identifier capable of surviving the reorganization of a Web site or the selling or transfer of intellectual property is a necessity in the electronic environment. Therefore, the aim of the AAP was to create an identifier for the electronic world that would represent "the ISBN of the 21st Century" (Bide 1997). The Digital Object Identifier (DOI) was the result of this initiative.

The DOI system can be defined as a "system for identifying and exchanging intellectual property in the digital environment" (http://www.doi.org). It fits within the Uniform Resource Name (URN) framework and takes the requirements of the URN as a fundamental starting point (Paskin 1999b; DOI Handbook 2002). In addition to identifying digital objects, the DOI system has the ability to provide a framework of maintenance for these intellectual property entities through the involvement of the International DOI Foundation (IDF). In 1998, the non-profit IDF was established to take over the development and management of the DOI framework. In addition to encouraging DOI developments, the IDF is also responsible for "influencing the development of standards that will ensure the appropriate level of value-added and quality control across the spectrum of participation" (Paskin 2000d). From the beginning, the role of the IDF was intended to be as promoter of the DOI, facilitating DOI registration and guiding institutions in the creation of applications to support the DOI (Paskin 1999a; Paskin 2000a). The first DOIs were registered directly through the IDF for a one-time fee of $1,000 per prefix and a minimal annual fee for each suffix. The IDF has begun to assume more of an administrative role, as originally intended and has begun designating other institutions as registration agencies for DOIs (DOI Handbook 2002). Thus, the business and economic models currently surrounding the DOI are expected to change drastically as each registration agency will be given free reign for choosing economic models for registering DOIs while the IDF remains responsible for policies and standards surrounding the DOI system.

There are a number of advantages that the DOI has to offer including persistency, extensibility, wide applicability and unlimited potential for future environments (Paskin 2000b). Persistency is a dominant characteristic of the DOI because regardless of URL location changes, the DOI will always remain the same. The information about the DOI is stored in a central DOI directory maintained by the IDF. If the location of the content were to change, the registrant of the DOI simply updates the URL location information in the directory so that it points to the new location. It is important to note that the persistency of the DOI is strongly dependent on how well registrants maintain the data within this central directory. Norman Paskin, IDF director, points out that "persistence is ultimately guaranteed by social infrastructure (policy); persistence is fundamentally due to people and technology can assist but not guarantee" (Paskin 2002). Without this social infrastructure supporting and maintaining the DOI system, there would be little hope for the ability of the DOI to remain persistent. If the DOI were not persistent, users may often see a screen such as that displayed in Figure 1 when attempting to resolve a DOI for which the content has been removed. While po-

FIGURE 1. Sample Error Page Displaying When an Entered DOI Does Not Resolve to Its Intended Article

tentially unavoidable, it is apparent that if the DOI is to truly remain persistent, these exceptions should be kept to a minimum.

The DOI system is also extensible as it allows and encourages further developments within its own framework. It can be applied to various manifestations of intellectual property, and could easily be implemented in future information environments (Paskin 2000b). In fact, examples of using DOIs to enhance electronic book features are just beginning to emerge (see DOI-EB examples at http://www.contentdirections.com). Given that the DOI is not limited to the framework of the Internet or other environments currently available, relatively few usage constraints are anticipated within future systems. The DOI is also considered to be an open system since anyone may build applications that utilize it.

Finally, because the use of a DOI must first be authorized by a registration agency, the DOI can be considered a fully managed system. This ensures that DOIs will only be assigned to documents that are regarded highly in need of persistent identification. Initially, the IDF was the major organization responsible for distributing and managing DOIs. However, as the use of the DOI has become more widespread, the IDF is beginning to assume more of a governing and delegation role, and has begun to assign the responsibilities of registering DOIs over to other institutions. Today there are at least five organizations designated by the IDF as DOI registration agencies. It is worth noting that although the IDF is a non-profit institution, registration agencies could be for-profit. CrossRef, a collaborative reference linking service for scholarly journal articles, was the first official registration agency appointed by the IDF to have the authority to register DOIs to CrossRef members and is discussed in more detail later in this paper. Other recent registration agencies include Content Directions, Learning Objects Networks, Enpia Systems in Korea, and Copyright Agency Limited in Australia.

## COMPONENTS OF THE DOI SYSTEM

The DOI system has a variety of components, including the DOI number, the mechanisms used to resolve the DOI, the metadata that accompanies each DOI, and various supporting policies.

### The DOI Number and Its Structure

A standard DOI is an alphanumeric string that consists of a prefix and a suffix. A common DOI would look something like **10.1000/182**, as the

DOI assigned to the HTML version of the IDF's DOI Handbook. The portion before the slash (10.1000) is called the prefix and contains two elements. The first element (10) is referred to as the registration agency prefix and represents the organization that is responsible for assigning the DOI prefix. Currently all DOIs begin with the number 10. This designates that the DOI is a handle assigned through the Corporation for National Research Initiatives (CNRI). The second element in the prefix (1000) is often referred to as the publisher prefix and is the number assigned to the organization registering the DOI (the registrant). This element has caused some degree of confusion in these early days of the DOI, as it seems as though the publisher responsible for the content displayed with the DOI could be identified by this element. However, it is not intended and is, in fact, discouraged that the prefix be used to represent publishers. The DOI is meant to transcend any exchange of ownership of intellectual property. While the second prefix element will always represent the original registrant, the clear distinction must be made between the organization that registered the DOI in the first place and the current owner of the intellectual property, as the two may certainly differ.

The portion of the DOI after the slash (in the example above, '182') is referred to as the suffix, and is chosen by the registrant of the DOI. A suffix can be any alphanumeric string that contains a combination of numbers and letters provided they are within the Unicode character set (http://www.unicode.org). The suffix is case sensitive and there is no restriction on length so long as the DOI has not been previously registered. Although the DOI is not intended to be a "smart" number, many organizations currently using DOIs have chosen to incorporate other identifiers that are already in existence (e.g., ISBN, SICI) and use them as the suffix. Further information about the specific syntax and composition required for the DOI can be found in the American National Standard developed by the National Information Standards Organization, ANSI/NISO Z39.84-2000.

### Resolution

Resolution can be described as the process of redirecting the user to the content identified by the DOI and is an essential component of the DOI system. In addition to the persistent identifier, the DOI System uses the CNRI Handle System for resolving its identifiers. The Handle System was created for "assigning, managing, and resolving persistent identifiers, known as 'handles,' for digital objects and other resources on the Internet" (http://www.handle.net). The DOI is only one type of handle currently being implemented through the overarching Handle System. Other institutions using handles supplied by the Handle System include the Library of Congress and the Defense Technical Information Center (DTIC).

There are currently two methods for resolving DOIs through the Handle System. The most widely used method today involves the resolution of DOIs through an HTTP proxy server. Under this method, a user clicks on a DOI link and the request is routed to the DOI directory through the DOI HTTP proxy server. The directory then looks up the DOI, identifies the URL that is on file and attempts to connect to the registered URL. For example, the DOI link <http://dx.doi.org/10.1000/182> resolves to the latest version of the DOI Handbook, currently located at the URL <http://www.doi.org/handbook_2000/index.html>. One of the limitations as currently implemented in the proxy server method is that it prevents the DOI from resolving to anything other than a single URL. In fact, the DOI standard does support resolving to multiple URLs using methods other than proxy server, so this is not an inherent problem.

The second method for resolving DOIs involves a plug-in resolver that enables Web browsers to recognize the handle protocol. This resolver can be downloaded free of charge from the Handle System's Web site (http://www.handle.net/resolver/index.html). Once installed, the plug-in works behind the scenes to recognize and resolve DOIs with the following structure: doi:10.1000/182. The advantages of the plug-in resolver over the proxy method include better performance and increased functionality (including multiple resolution capabilities). However, it cannot be expected that every user accessing electronic journal articles would be willing to go through the process of locating and downloading the plug-in. Although the HTTP proxy resolution is the more popular of the two resolution methods presented above, users should be aware that this method could be replaced by better functionality in the future. In fact, some individuals hope that one day native resolution of handles may be built into future versions of Web browsers (Paskin 2000d).

One of the current developments for the DOI system includes implementing multiple resolution capabilities. Because there can be many different instances of an intellectual property (i.e., an electronic journal could be presented in PDF or HTML formats), it is important to allow ways for users to access these multiple instances. Multiple resolution provides a solution for allowing the DOI to locate different formats, including abstracts, full-text in PDF or HTML formats, etc. The underlying technology of the Handle System already has the capability of performing multiple resolution, and the IDF has stated its desire to implement this functionality in the near future (Paskin 2000d). A demonstration of the multiple resolution functionality of the Handle System is available at the Content Directions Web site (http://www.contentdirections.com). Additional descriptions of multiple resolution capabilities of the DOI System can be found at <http://www.doi.org/mult_resoln.radextraParse. html>. Although current developments for multiple resolution are still in the pipeline, some believe that these efforts must come to fruition in order to utilize the full potential of the DOI system (Paskin 1999a; Paskin 1999b; Beit-Arie et al. 2001). However, many publishers prefer to have control over the resolution capabilities and tend to resolve their DOIs to a single URL that then provides multiple options for the user, as identified by the DOI resolution for an American Chemical Society article shown in Figure 2.

## *Metadata*

While the DOI number is important for identifying objects, a description of the content is still necessary to fully capitalize on the DOI System (ANSI/NISO Z39.84-2000). This requires that well formed metadata about the object be defined upon registration with the IDF. The IDF has thus adopted a set of metadata principles from the Interoperability of Data in E-Commerce Systems (INDECS) project that prescribe the following:

- Unique identification
- Functional granularity
- Designated authority
- Appropriate access

The unique identification principle states that "every entity should be uniquely identified within an identified namespace" (Rust and Bide 2000).

### FIGURE 2. Sample DOI Resolution

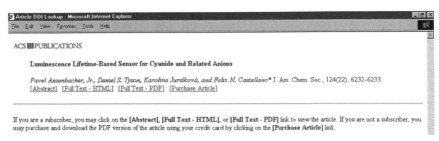

The DOI clearly satisfies this principle since it must be unique by definition. The second principle of functional granularity can be stated simply as identifying an object only when the need arises. This is satisfied by the DOI system through the flexibility of DOIs since they can be assigned to virtually any aspect of the intellectual property (i.e., abstract, PDF, etc.). The principle of designated authority states that "the author of an item of metadata should be securely identified" (Rust and Bide 2000). In other words, the metadata must be supplied by an authoritative source (in most cases, the registrant of the DOI) in order to ascertain that the description is in fact reliable. Finally, the principle of appropriate access ensures that those who depend on the metadata will indeed have access to it (Paskin 1999a).

The DOI is not limited to digital objects alone, but also provides a level of flexibility to be applied outside of the digital realm to other manifestations of a material. For example, DOIs could be used to identify books (electronic or print), multimedia materials, or any other types of documents that may be created in the future. Given this flexibility, the DOI System must reserve a level of structure and specificity to avoid hindering itself with a "crippling ambiguity" (Paskin 1999a). This point only emphasizes the importance of having a consistent method for the creation and provision of DOI metadata. To achieve this goal, the DOI System has constructed a minimal set of "kernel metadata" that must be declared at the time a DOI is registered. This kernel metadata consists of the following mandatory elements (DOI Handbook 2002):

- DOI
- Title
- Type
- Mode
- Agent's role
- Primary agent
- DOI Application Profile

DOI Application Profile(s) (DOI-AP) contain additional metadata about the content identified by the DOI and are assigned at the time a DOI is registered. It should be noted that DOI Application Profiles are still in the process of being defined and will differ according to the registration agency responsible for assigning the DOI. The DOI-AP is intended to be the "functional specification of an application of the DOI to a class of intellectual property entities that share a common set of attributes" (DOI Handbook 2002). It is intended that each DOI be assigned to at least one

DOI-AP. However, more than one DOI-AP can be assigned provided that the DOI adheres to the various rules corresponding to each specific DOI-AP. The required metadata within the CrossRef database is actually quite minimal. A sample XML-DTD displaying the CrossRef metadata for an American Institute of Physics (AIP) article is shown in Figure 3.

## Policies

The careful creation of policies and standards for the DOI System will ensure that "publishers will dance to the same rhythm" (Rosenblatt 1997). The International DOI Foundation (IDF) is responsible for for-

FIGURE 3. Sample METADATA in XML-DTD Form

```
File   Edit   View   Favorites   Tools   Help

<?xml version="1.0" encoding="UTF-8" ?>
<!DOCTYPE doi_batch (View Source for full doctype...)>
- <doi_batch version="0.3">
  - <head>
      <doi_batch_id>XRef::Resolver_30-May-2002@16:45:20</doi_batch_id>
      <timestamp>30-May-2002@16:45:20</timestamp>
    - <depositor>
        <name />
        <email_address />
      </depositor>
      <registrant />
    </head>
  - <body>
    - <doi_record type="full_text">
      - <doi_data>
          <doi>10.1063/1.1462847</doi>
          <url>http://dx.doi.org/10.1063/1.1462847</url>
        </doi_data>
      - <journal_article_metadata>
        - <article>
          - <enumeration>
              <volume>91</volume>
              <first_page>5158</first_page>
            </enumeration>
          </article>
        - <journal>
            <full_title />
            <issn type="print">0021-8979</issn>
          </journal>
        </journal_article_metadata>
      </doi_record>
    </body>
</doi_batch>
```

mulating policies that guarantee the DOI system "behaves in ways that are predictable and consistent" (DOI Handbook 2002). As Mark Bide states in his report, *In Search of the Unicorn,* the "potential value of the DOI (for rights owners, users and intermediaries) can only be realized within a framework of standards" (Bide 1998). Although board members of the IDF are ultimately responsible for making policy decisions, all IDF members remain closely involved in the policy development process. A few of the policies already formulated include the following:

- A DOI can be used to identify any intellectual property entity.
- The primary focus of the DOI is on the management of intellectual property entities.
- All DOIs must be registered with a global DOI registry.
- Minimum kernel metadata will be declared for all DOIs registered.

Currently the IDF is also working to formulate policies regarding the roles and relationships between itself and registration agencies. Given the development of these four components of the DOI System, the next logical step necessary was to determine what applications the DOI might have. The first significant application surfaced within the publishing realm and involves linking references within scholarly journals.

## REFERENCE LINKING AND CrossRef

Reference linking has been defined as "the ability to go directly from a citation to the work cited, or to additional information about the cited work" (Caplan 2001a) and is a fairly recent addition to electronic journal articles. Although the electronic environment provides new and promising opportunities for reference linking, moving between scholarly documents is certainly not a new concept for scholars. In fact, the very process of scientific communication is "founded on dependable links between articles" (Paskin 2000c). The process of retrieving these articles will certainly be expedited in the electronic environment and there is no doubt that the introduction of reference linking has added tremendous value to scholarly communication (Brand 2001; Caplan 2001a).

As full-text electronic journals began to proliferate in the mid-1990s, publishers began experimenting with the provision of links from references within their articles to other articles both within and outside their journals. However, it is rarely the case that the same publisher publishes all references listed for a given work. Therefore, publishers began to make external agreements with other publishers for permission to link to their journal articles. Many publishers developed their own custom reference linking technologies. However, agreement restrictions allowing only the provision of links to publications owned by one company or to a few external publishing companies seemed to leave the reference linking service lacking. After all, scholars do not care much which publishers deliver the desired content contained in an article, so long as they are in fact able to obtain the article (Brand 2001; Paskin 2000c). Linking across publishers was proving to be a difficult task for any one publisher to tackle. These issues led to various conversations among publishers concerning the standardization of reference linking. This prompted the development of a prototype reference-linking project in 1999 referred to as DOI-X. The purpose of the DOI-X prototype was to "address the integration of metadata registration and maintenance with basic DOI registration and maintenance" (Atkins et al. 2000). It quickly became recognized that combining the persistency of the DOI with the added value of reference linking to provide persistent links across publishers could prove to be a very powerful application of the DOI (Atkins et al. 2000; Paskin 1999a) and CrossRef was born out of this initiative.

CrossRef is a collaborative reference linking service operated by the Publishers International Linking Association (PILA). CrossRef was formed through the collaboration of a number of leading STM publishers in 1999 and was the first registration agency designated by the IDF for the distribution of DOIs. It is "an infrastructure for linking citations across publishers" with a mission to "serve as the complete citation linking backbone for all scholarly literature online, as a means of lowering barriers to content discovery and access for the researcher" (http://www.crossref.org). As stated above, the CrossRef protocols were first tested in 1999 in the prototype reference linking system known as DOI-X (Atkins et al. 2000). Although initially developed by STM primary publishers, the services provided by CrossRef were soon considered an important asset for all areas of scholarly communication and now involve primary and secondary publishers in all disciplines, librarians, and scholars alike. Today, there are over one hundred publishers, spanning all areas of scholarly publishing, as well as over thirty library affiliates participating in the CrossRef initiative. The service for looking up DOIs and metadata provided by CrossRef is not a service open to all users. It is designed as a service for organizations interested in locating and using DOIs in reference linking. Once a publisher obtains membership to CrossRef, it is allowed access to the CrossRef database and may send reference citations from each journal article to be linked to a "reference resolver." This reference resolver is the component that allows the registrant to retrieve DOIs and thus, create links for other publications provided within the CrossRef database. Ed Pentz, executive director of CrossRef, makes the argument that CrossRef is able to provide "the 'missing link' in linking, making broad-based linking efficient and manageable for large and small publishers" (Pentz 2001a). The actual format the publishers use to display the links made through CrossRef vary considerably.

For example, Springer uses icons to designate each link, as displayed in Figure 4, while others simply use the standard hyperlink. It has been stated that the "DOI/CrossRef infrastructure provides a very powerful and generalized mechanism for linking to electronic journal articles across a large number of independent and heterogeneous publisher systems" (Beit-Arie et al. 2001). Due to the success of linking journal article references (and because journal articles are not the only mechanism for scholarly communication), CrossRef is currently expanding the ap-

FIGURE 4. Sample References Page

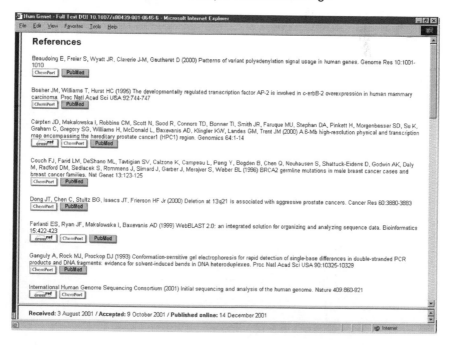

plication of DOIs to include other methods of publication including conference proceedings, preprints, book chapters, etc. (Paskin 2002). Reference linking with DOIs has made considerable progress within the last few years. However, it is not without its problems and challenges. Perhaps the problem that raises most concern from libraries is the "appropriate copy" problem and is discussed in more depth in the next section.

Another benefit of reference linking is the capability of providing links not only for articles that are referenced within the article, but also providing access to the articles that have later cited the given article. For example, the American Institute of Physics (AIP) provides a section at the bottom of the abstract and references list called "Citing articles," as shown in Figure 5. It appears as though AIP is currently only providing links to articles that they publish. However, bi-directional linking of citations is an important feature for increasing the value of linking scholarly works using the DOI.

## THE "APPROPRIATE COPY" PROBLEM

Although reference linking with DOIs certainly adds value to the electronic publishing process, it has also presented a major problem to information providers today known as the "appropriate copy" problem (Caplan and Flecker 1999; Beit-Arie et al. 2001). A scenario demon-

FIGURE 5. Sample of Citing Articles Now Included in Abstract and HTML Full-Text Versions of AIP Articles

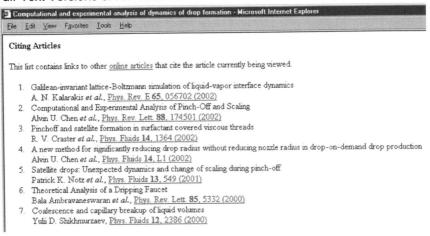

strating the appropriate copy problem follows: John, a graduate student at University XYZ, is reading a research article online provided by Publisher A. John finds a citation at the end of the article of particular interest to him, so he clicks the link to the reference. However, instead of being directed to the article, a page from Publisher B (owner of the cited article) is displayed stating John must purchase the article in order to obtain access to it. What John would probably not realize, is that the library at University XYZ does indeed have access to the journal through alternate methods. It could be that the journal has been loaded locally, is available in print, or is provided through an aggregator service.

One solution to this problem comes in determining a way in which the resolver would look first for resources already accessible locally or through subscription. If neither of these options is available, then alternatives that require payment can be presented. Some of the issues surrounding the appropriate copy problem will require consideration at an institutional level, based on different network structures and large variation in subscriptions across institutions (Beit-Arie et al. 2001). The recent development of the OpenURL framework holds promising potential for addressing the appropriate copy problem. The OpenURL is intended to "provide a standardized format for transporting bibliographic metadata about objects between information services" (Van de Sompel and Beit-Arie 2001). Last year, a prototype project striving to address the appropriate copy problem utilized the OpenURL framework and is documented further in Beit-Arie et al. (2001). The results of this project were highly satisfactory, demonstrating that it is indeed possible to implement a system that is capable of local resolution. The most significant aspect of the project was the success with which the high level of collaboration among various institutions was carried out (Caplan 2001b). However, it is important that libraries and other information providers continue to be involved in the development of the DOI and the reference linking schemes. Librarians must "view reference linking not as an external development, but rather as an evolving set of organizational and economic options that they can and should help influence" (Caplan 2001a).

## *SCI-TECH PUBLISHERS AND THE DOI*

CrossRef has now been in place for over two years, having grown from twelve founding publishers in the year 2000 to over one hundred publishers and over thirty library affiliates today. As we have seen, major initiatives for providing effective reference linking have also been in development. This section focuses on how ten different STM publishers are currently utilizing the DOI within their electronic journals. It should be noted that these findings are based on a preliminary assessment performed on a number of journals provided by each publisher and are by no means intended as a comprehensive analysis of all journals for a given publisher. Additionally, since the resources made available in the electronic environment is constantly changing and being updated, data presented in the analysis may very well become outdated before this article is published (e.g., dates of electronic journal availability).

There are three distinct aspects that must be taken into consideration when discussing the inclusion of DOIs in electronic publications. The first question to be addressed is whether or not DOIs have been assigned to the publications. Second, if DOIs have been assigned, are they accessible (visible) on the publication itself? Third, are publishers utilizing the DOI within reference linking schemes for their publications?

### *DOI Assignment*

As electronic journals can still be considered a fairly recent addition to the world of publishing, the actual number of electronic journals, and dates for which the journals are available, varies considerably. Some STM publishers, like the Association of Computing Machinery (ACM) and American Physical Society (APS), provide fairly extensive electronic back files to many of their journals, including those as far back as 1940 and 1950. Likewise, the assignment of DOIs to electronic publications varies not only between publishers, but also within specific journals. For example, the ACM has assigned DOIs to almost all of their publications available online, including not only journal articles, but also conference proceedings, transactions, etc. However, APS currently appears to have DOIs assigned only to articles from January 2000 forward. Data for the assignment of DOIs is documented in Table 1. It should be noted that, for the purposes of this assessment, the dates for articles with DOIs is primarily based on visible indications of the DOI on a version of the publication (usually the abstract). Some publishers, such as the Institute of Electrical and Electronics Engineers (IEEE), have assigned DOIs to their documents without making the number accessible from the publication. Others (e.g., Elsevier, Springer) appear to have registered DOIs for most journal articles, without explicitly displaying the DOI on earlier publications. Materials without a visible DOI on the publication have not been taken into account for the dates of included DOIs in Table 1.

## TABLE 1. Assignment of DOIs

| Publisher | Dates of Online Articles | Dates with DOIs displayed* | DOIs displayed for all articles? |
|---|---|---|---|
| Academic Press | varies (most 1993- ) | varies | Y |
| ACM | 1954- | 1954- | Y |
| ACS | Sept 1981- (PDF)<br>Jan 1996- (HTML) | 1996- | N |
| AIP | 1975-1985 (abstracts only)<br>1985- | 1975-1985[†]<br>1985- | Y |
| APS | 1998-[‡] | Jan 2000-[‡] | N |
| Elsevier | varies | June 2001-[§] | N |
| IEEE | varies | unknown[§] | unknown |
| Springer | varies (mid-1990s) | varies[§] | N |
| Wiley | 1996- | 1996- | Y |
| World Scientific | 1997- | 2002- | N |

*DOIs may be assigned for additional articles but not displayed.
[†]DOIs have been assigned for all articles and abstracts, however abstract DOIs don't appear to resolve.
[‡]Articles from the Physical Review Online Archive are also available electronically, with assigned DOIs.
[§]These publishers all have assigned DOIs for additional articles that do not display the DOI.

## Displaying the DOI

Publishers have started to display DOIs within various formats of their publications. As displayed in Table 2, four of the ten publishers are including the DOI within the table of contents, before a user even accesses a specific article. For example, the DOIs presented in the table of contents for journals of the American Chemical Society (ACS) are actually hyperlinks to the HTTP proxy resolution of the DOI.

Nine of the ten publishers reviewed have listed the DOI somewhere on the abstract page for the most recent articles, and all publishers who provide an HTML version of the document also list the DOI on the HTML document. Most publishers (with the exception of Elsevier and ACM) are now providing DOIs on the print versions of articles as well. If the DOI is intended to be useful to the scholarly community and to be truly extensible beyond Internet applications, it is crucial that publishers consider how to facilitate user access to the DOI. To some degree it may not be considered necessary for the user to have access to the DOI, especially when one considers its length, arbitrary string of numbers and overall confusing nature. However, the counter-argument could be made that if a user would like to bookmark a particular URL for future reference or include it in his/her references list but did not have the DOI, he/she would lose the content of the material if its location changed.

TABLE 2. Location of DOIs on Publications

| Publisher | TOC | Abstract | PDF | HTML | Print/Dates |
|---|---|---|---|---|---|
| Academic Press | Y | Y | Y | N/A | Y-Jan 2000 |
| ACM | N | Y | | N/A | N |
| ACS | Y | Y | Y | Y | Y–Nov 1998 |
| AIP | N | Y | Y | N/A | Y |
| APS | N | Y | Y | N/A | Y |
| Elsevier | N | Y | N | Y | N |
| IEEE | N | N | N | N | N |
| Springer | Y/N | Y | Y | Y | Y |
| Wiley | Y | Y | N | Y | Y-2002 |
| World Scientific | N | Y | N | N/A | N |

## Reference Linking and the DOI

Since all ten STM publishers reviewed for this assessment are members of CrossRef (all but two are founding members), it was predicted before the assessment began that most, if not all, publishers would provide reference linking capabilities within their electronic journals. However, the findings indicate that only seven are currently providing links to references within their articles, as presented in Table 3.

The three publishers who do not provide reference linking within their articles only provide PDF versions of their publications, which could explain the lack of reference linking. However, a few other publishers (e.g., Wiley, APS and AIP) are working to remedy this situation by providing "extended abstracts" in HTML that include lists of references, thus allowing reference linking capabilities to be provided. Of the seven who do link their references, they are using DOIs in a variety of ways. For instance, the American Chemical Society (ACS) does not appear to be using DOIs for any references. Reference linking to relevant articles is provided through ChemPort, a reference linking service provided by Chemical Abstracts. From the references page, ACS articles are linked with a URL that does not include the DOI. However, after clicking the link to a given article, one can determine that ACS is actually using a URL other than the proxy server to access the DOI (e.g., http://pubs3.acs.org/acs/journals/doilookup?in_doi=10.1021/ac950935b). They are undoubtedly redirecting to the DOI through the proxy server from their local URL. Springer, on the other hand, provides icons (where appropriate) for ChemPort, PubMed, CrossRef and Springer.Link. Both the CrossRef and Springer.Link icons link the user to the given article

## TABLE 3. Reference Linking and DOIs

| Publisher | Linked References Available | Use of DOIs for other publishers | Use of DOIs for own articles |
|---|---|---|---|
| Academic Press | Y | Y | N* |
| ACM | N | N/A | N/A |
| ACS | Y | N | N |
| AIP | Y | Y | N* |
| APS | Y | Y | Y* |
| Elsevier | Y | Y | N |
| IEEE | N | N/A | N/A |
| Springer | Y | Y/N† | Y |
| Wiley | Y | N | N |
| World Scientific | N | N/A | N/A |

*These publishers use own URL linking schemes to link to their journals (e.g., Academic Press uses URLs like the following: http://www.idealibrary.com/links/doi/10.1016/mchj.1995.1105 while AIP uses the following: http://link.aps.org/abstract/PRE/v65/e056702).
†Access to DOIs in references depends heavily on the journal.

using the DOI proxy server URL. Wiley, however, provides a link entitled 'Links' for all references within the article, regardless of whether links are available or not. Many references link to PubMed or ISI bibliographic citations, others link to outside publishers. However, based on this assessment, it does not appear that Wiley is using the DOI, unless, of course, it is being used behind the scenes.

Overall, the strongest conclusion that can be drawn from this initial assessment is that publishers are currently transitioning into the use of the DOI in their electronic publications. The DOI has begun to make an entrance into many reference linking schemes. However, many publishers are still maintaining their own custom linking systems for their publications, as demonstrated by ACS, AIP and Springer. It makes sense that publishers would want to maintain the links to their articles on their own servers. After all, the links are, in most cases, already established, the local resolution time could potentially be faster than other methods, and since the DOIs are registered, users could still retrieve the full-text through the DOI proxy server if needed. As stated previously, the IDF does not consider the proxy server method as the long-term resolution method of choice and should not be considered the only reliable method of resolving DOIs. However, the problem with the current approach many publishers are taking is the lack of consistency within their system for the utilization of DOIs. Another point that has not yet been documented in the literature, is how linked references that are updated as DOIs are made accessible to back files. Are publishers responsible for re-running their lists of references through the reference resolver in order to obtain DOIs for these 'newer' materials? This is a problem that should be addressed soon in order to prevent a large number of references lacking useful and relevant DOIs.

## CONCLUSION

Solutions to the issues of appropriate copy and intellectual property management are large tasks and will not and should not be resolved in a hasty manner. In addition, adequate solutions will only be obtained if various communities collaborate to identify a solution that satisfies all parties involved (Caplan and Flecker 1999). As Van de Sompel and Hochstenbach state "the creation of a fully interlinked information environment . . . would require either an information monopoly or extensive partnerships" (Van de Sompel and Hochstenbach 1999). As it is, the DOI system implemented by CrossRef supports a publisher centric scholarly communication system built through partnerships.

The DOI certainly has potential to become a strong identifier for providing persistent access to electronically published literature. As we have seen, the system is not flawless and it will continue to coexist with other identifiers (e.g., URLs, etc.) in the electronic realm. In order to allow the DOI System to develop to its fullest potential, it is mandatory to have the highest level of collaboration from various organizations (Caplan and Flecker 1999). These levels of collaboration are not only necessary but are entirely feasible and hopefully, the "end results of this collaboration will be to make robust, open reference linking possible across all legal copies of articles, which will serve the best interests of scholars, libraries and publishers" (Beit-Arie et al. 2001). Despite its short history, the DOI system has made considerable progress for identifying and managing access to digital objects, particularly within and between scholarly journals. Although the official responsibility of developing the DOI system policies falls solely with the IDF, this does not mean that other institutions should not be involved in the process. Librarians, secondary publishers, and other information providers must insist on their involvement, make themselves aware of new developments, and voice their own needs and desires regarding the development of the DOI System. Now is the time to take action.

# REFERENCES

American National Standards Institute (ANSI). 2000. *Syntax for the Digital Object Identifier*. Bethesda: NISO Press. ANSI/NISO Z39.84-2000. http://www.niso.org/standards/resources/Z39-84-2000.pdf (Accessed 22 May 2002).

Atkins, Helen, Catherine Lyons, Howard Ratner, Carol Risher, Chris Shillum, David Sidman, and Andrew Stevens. 2000. Reference linking with DOIs: A case study. *D-Lib Magazine*, 6 (2), (February). http://dx.doi.org/10.1045/february2000-risher (Accessed 22 May 2002).

Beit-Arie, Oren, Miriam Blake, Priscilla Caplan, Dale Flecker, Tim Ingoldsby, Laurence W. Lannom, William H. Mischo, Edward Pentz, Sally Rogers, and Herbert Van de Sompel. 2001. Linking to the appropriate copy: Report of a DOI-based prototype. *D-Lib Magazine*, 7 (9). http://dx.doi.org/10.1045/september200-caplan (Accessed 21 May 2002).

Bide, Mark. 1998. "In search of the unicorn: The Digital Object Identifier from a user perspective." *BNBRF Report 89*, Book Industry Communication, (February). http://www.bic.org.uk/unicorn2.pdf (Accessed 22 May 2002).

Brand, Amy. 2001. CrossRef turns one. *D-Lib Magazine*, 7 (5), (May). http://dx.doi.org/10.1045/may2001-brand (Accessed 22 May 2002).

Caplan, Priscilla. 2001a. Reference linking for journal articles: Promise, progress and perils. *portal: Libraries and the Academy*, 1.3: 351-356. http://muse.jhu.edu/journals/portal_libraries_and_the_academy/v001/1.3caplan.html (Accessed 23 May 2002).

_____2001b. A lesson in linking. *Library Journal Net Connect*, 16-18, Fall 2001.

Caplan, Priscilla and Dale Flecker. 1999. *Choosing the appropriate copy*. Digital Library Federation Architecture Committee. (September). http://www.niso.org/news/reports/DLFarch.html (Accessed 25 May 2002).

Caplan, Priscilla and William Arms. 1999. Reference linking for journal articles. *D-Lib Magazine*, 5 (7-8), (July/August). http://dx.doi.org/10.1045/july99-caplan (Accessed 22 May 2002).

Content Directions, Inc. http://www.contentdirections.com (Accessed 22 May 2002).

*CrossRef*. 2000. Publishers International Linking Association, Inc. http://www.crossref.org (Accessed 22 May 2002).

Davidson, Lloyd A. and Kimberly Douglas. 1998. Promise and problems for scholarly publishing. *The Journal of Electronic Publishing*. 4 (2), (December). http://www.press.umich.edu/jep/04-02/davidson.html (Accessed 22 May 2002).

DOI Handbook. 2002. Version 2.1.0. Oxford: International DOI Foundation. Available: http://dx.doi.org/10.1000/182 (Accessed 22 May 2002).

Grogg, Jill E. 2002. Thinking about reference linking. *Searcher*, 10 (4). http://www.infotoday.com/searcher/apr02/grogg.htm (Accessed 3 May 2002).

Handle System. 2000. Corporation for National Research Initiatives. (11 April) http://www.handle.net (Accessed 22 May 2002).

<indecs>. 2001. http://www.indecs.org (Accessed 22 May 2002).

International DOI Foundation. http://www.doi.org (Accessed 22 May 2002).

Lynch, Clifford. 1997. Identifiers and their role in networked information applications. *ARL: A Bimonthly Newsletter of Research Library Issues and Actions* 194 (October

1997). Available: http://www.arl.org/newsltr/194/identifier.html (Accessed 22 May 2002).

Paskin, Norman. 1999a. DOI: Current status and outlook. *D-Lib Magazine* 5 (5), (May). http://dx.doi.org/10.1045/may99-paskin (Accessed 22 May 2002).

_____1999b. DOI: An overview of current status and outlook. *ICSTI Forum*, no. 30. http://www.icsti.org/forum/30/#paskin (Accessed 25 May 2001).

2000a. DOI discussion paper: DOI deployment. Version 2.0, (February). http://www.doi.org/deployment2.pdf (Accessed 14 February 2002).

_____2000b. Digital Object Identifier: Implementing a standard digital identifier as the key to effective digital rights management. (April). http://dx.doi.org/10.1000/174 (Accessed 22 May 2002).

_____2000c. E-citations: actionable identifiers and scholarly referencing. *Learned Publishing* 13 (3), (July). http://dx.doi.org/10.1087/09531510050145308 (Accessed 22 May 2002).

_____2000d. *'From one to many': the next stage in the development of DOI functionality.* Version 1.0, (August). http://dx.doi.org/10.1000/190 (Accessed 22 May 2002).

_____ 2002. Digital Object Identifiers. ICSTI Seminar: Digital Preservation of the Record of Science (Feb 14/15 2002), to be published by IOS Press.

Paskin, Norman, Eamonn Neylon, Tony Hammond and Sam Sun. 2002. Uniform Resource Identifier (URI) scheme for Digital Object Identifiers (DOIs). Work in progress.

Pentz, Ed. 2001a. CrossRef: The missing link. *College and Research Libraries News* 62 (2), (February).

_____ 2001b. CrossRef: A collaborative linking network. *Issues in Science and Technology Librarianship* 29. (Winter). http://www.library.ucsb.edu/istl/01-winter/article1.html (Accessed 1 May 2002).

Reid, Calvin. 2001. Selling books with the DOI. *Publishers Weekly* 248 (43), (October 22): 28.

Rosenblatt, Bill. 1997. Solving the dilemma of copyright protection online. *The Journal of Electronic Publishing*, 3 (2), (December). http://www.press.umich.edu/jep/03-02/doi.html (Accessed 22 May 2002).

Rust, Godfrey and Mark Bide. 2000. *The <indecs> metadata framework: Principles, model and data dictionary.* Version 2, (June*).* http://www.indecs.org/pdf/framework.pdf (Accessed 22 May 2002).

Software & Information Industry Association. 2001. The digital object identifier: The keystone for digital rights management. http://www.siia.net/divisions/content/doi.pdf (Accessed 22 May 2002).

Van de Sompel, Herbert and Patrick Hochstenbach. 1999. Reference linking in a hybrid library environment. Part 1: Frameworks for linking. *D-Lib Magazine*, 5 (4), (April). http://dx.doi.org/10.1045/april99-van_de_sompel-pt1 (Accessed 22 May 2002).

Van de Sompel, Herbert and Oren Beit-Arie. 2001. Open linking in the scholarly information environment using the OpenURL framework. *D-Lib Magazine*, 7 (3), (March). http://dx.doi.org/10.1045/march2001-vandesompel (Accessed 3 May 2002).

# When Vendor Statistics Are Not Enough: Determining Use of Electronic Databases

Amy S. Van Epps

**SUMMARY.** Many libraries have large collections of electronically available databases including journal article and conference paper indexes, full-text vendor catalogs, and standards databases. Which of these resources are being used and to what level becomes a point of interest. A quick re-direct Web-log has been created to track the number of times a particular link is selected, providing a consistent comparison of different resources. The resulting information can be used to determine if what the library provided is being used and if it can be marketed more effectively, which, ultimately, will aid in a cost/benefit analysis for budget decisions. *[Article copies available for a fee from The Haworth Document Delivery Service: 1-800-HAWORTH. E-mail address: <docdelivery@ haworthpress.com> Website: <http://www.HaworthPress.com> © 2001 by The Haworth Press, Inc. All rights reserved.]*

**KEYWORDS.** Database statistics, scientific and technical libraries statistics, database use, collection development statistics

Amy S. Van Epps is currently Assistant Engineering Librarian and Assistant Professor of Library Science, Purdue University, and maintains an active membership in the American Society for Engineering Education, Engineering Libraries Division. She holds a Bachelor of Arts in Mechanical Engineering, an MSLS, and a Master of Engineering in Industrial and Management Engineering.

[Haworth co-indexing entry note]: "When Vendor Statistics Are Not Enough: Determining Use of Electronic Databases." Van Epps, Amy S. Co-published simultaneously in *Science & Technology Libraries* (The Haworth Information Press, an imprint of The Haworth Press, Inc.) Vol. 21, No. 1/2, 2001, pp. 119-126; and: *Information Practice in Science and Technology: Evolving Challenges and New Directions* (ed: Mary C. Schlembach) The Haworth Information Press, an imprint of The Haworth Press, Inc., 2001, pp. 119-126. Single or multiple copies of this article are available for a fee from The Haworth Document Delivery Service [1-800-HAWORTH, 9:00 a.m. - 5:00 p.m. (EST). E-mail address: docdelivery@haworthpress.com].

10.1300/J122v21n01_10

As library resources for all disciplines become increasingly available in electronic format, libraries are faced with many new issues. Among these are changing demands for library and information instruction to students, marketing of library services and determining which of the many electronic resources our clientele is using.

Being able to determine which resources are being used, and at what level, is the first step in addressing questions such as: Are the electronic materials the library supplies being used? Is the library providing access to what is needed? Do some resources need to be promoted more vigorously? Additionally, librarians need to know who is using the resources, and if users are connecting from an on-campus computer in a given school or department or from a computer in an off-campus location. As Dowling states, librarians need to quantify information since the ability to do so may determine continued funding for the resources (Dowling 2001). In times of tight budgets, libraries are often faced, at best, with flat materials budgets while prices for journals and online resources continue to rise at percentages larger than inflation. Use data should help determine which items are most critical to maintain. At first consideration, the fact that these resources are being supplied and used online would lead one to believe that gathering and analyzing data on which resources are being used most frequently would be relatively simple.

Unfortunately a number of obstacles arise in this seemingly simple task. In many instances, vendors of the electronic products provide statistics on use but only at their own discretion. Multiple vendors lead to multiple formats and different statistical data. If a library receives all of its electronic materials through one vendor, then it is likely to have very few problems, but most science and technology libraries do not have that option. In order to provide all the critical resources, libraries must use a variety of vendors and the statistics from those various vendors are not consistent or easily reconciled with each other, or in some cases, are non-existent. One vendor may report the number of searches performed, another the number of sets returned, a third the number of records retrieved and/or the number of records viewed, while a fourth may provide the number of logins to the file. Furthermore, when one vendor provides multiple files, e.g., Cambridge Scientific Abstracts (CSA), the number of logins to the vendor service is provided, and each file provided by that vendor shows the number of queries performed. Even if two vendors report the number of searches, the data may not be comparable, as each vendor may have a different definition of what constitutes a search and when a new one begins. Covey identifies incompatibility of data along with multiple formats, delivery methods and schedules for providing data, and the lack of intelligible data as usual complaints with vendor supplied statistics (Covey 2002).

Several organizations are formulating standards for vendor reporting of electronic product statistics. A first stop toward addressing these problems was taken by the International Coalition for Library Consortia (ICOLC), which wrote guidelines for the minimum information that vendors should be supplying to libraries (ICOLC 1998). The guidelines include a list of expected data elements including: number of queries (and a definition of a query), number of session/logins as a measure of simultaneous use, and the number of turnaways, if applicable. More recently, the Association of Research Libraries (ARL) is sponsoring an E-Metrics study on developing statistics and performance measures for electronic materials. Phase One of this project identified current practices for statistics and performance measures in ARL libraries using surveys and site visits, and organized a group to begin talking with vendors about statistics (Shim 2000). Phase Two of the ARL E-metrics project has gone on to define a number of recommended statistics for library networked resources and assist in defining how these statistics could be used (Shim 2001). The goal of these groups is ultimately to gain agreement and compliance by the vendors on a consistent set of statistics, thus relieving libraries from gathering and analyzing their own data to gain the information needed to make decisions, and to assist libraries in putting their statistical information to use.

The problem of irreconcilable statistics generated the current project at Purdue University. Like several of the libraries in the ARL study, Purdue needs a concrete indication of the most used resources. While the library receives information from most vendors and generates numbers for items loaded locally, the numbers provided are most helpful in tracking use trends for a particular file, not in comparing the resources to each other.

A pilot project of gathering statistics was started to provide numbers that, while not perfect, can be compared to each other in a meaningful fashion and provide data on which resources are being used. The information is created by placing a redirect script call before the URL for each resource. The information gathered is sometimes referred to as a 'click-through.' The redirect CGI script writes a line to a log file that records which resource link was followed before passing the user along to the requested resource. This log files is then analyzed to create the information regarding which of the files are used most often.

Transaction log analysis, which is the technology being used, is by no means new (it has been in existence for about 25 years), nor is Purdue the first library to apply it (Covey 2002). Transaction log analysis is the process used by most library management systems to determine the types of searches performed most often in the library catalog and other electronic resources. The ARL E-metrics project visited the University of Pennsylvania libraries and learned they use a similar system to track what they call 'attempted logins' and at Texas A&M a click-through page is displayed when a user selects an electronic journal from the library catalog to count the number of times the journal is used (Shim 2001; Burford 2001). Using an actual intermediary page may be helpful to display license or copyright agreements for electronic journals, but adds an unnecessary step if your goal is to count click-throughs. Scripts, like the one in Figure 1, provided the ability to track use by writing to the log and automatically redirecting the user to the resource requested.

FIGURE 1. Redirect Logging Script

```
$curlog = "redirect-cgi";
$delim = "\n";
$field_sep = "\t";

($sec, $min, $hour, $mday, $mon, $year) = localtime( time );
$mday = '0' . $mday if (length( $mday ) < 2);
$TimeOnly = sprintf("%02d:%02d:%02d", $hour, $min, $sec);
$month = (Jan,Feb,Mar,Apr,May,Jun,Jul,Aug,Sep,Oct,Nov,Dec)[$mon];
$yr = 1900+$year;
$DateOnly = $mday."/".$month."/".$yr;
$tzone = "-0500";
$logdate ="[".$DateOnly.":".$TimeOnly." ".$tzone."]";
&re_direct( "$ENV{'QUERY_STRING'}" );
open( FILE, ">>$curlog" );
print FILE "$ENV{'REMOTE_ADDR'} - $ENV{'HTTP_REFERER'} - logdate
\"GET".$ENV
{'QUERY_STRING'} . " HTTP/1.0\" 301
1\n";
close( FILE );
sub re_direct {
local ($location) = @_;
print <<"--end--";
Content-type: text/html
Location: $location

<h1>301 Redirect</h1>
Document is located at <a href="$location">$location</a>
--end--
}
```

There are several shortcomings to this type of data collection. First, since it is still in a limited implementation at Purdue, we are not tracking all the use of our files, only those uses that are originating from the Engineering Library's Web page. A comparison of the logged redirect numbers to the total number of institutional logins supplied to us by the vendor indicates a large number of uses are originating from pages other than on the Engineering Library's Web site. That is, a large number of users have bookmarked the vendor Web site and are going directly to the site or starting from a page other than the Engineering Library Web page. Second, there is no guarantee that each click-through is an actual use of the database. For example, when a librarian is showing a patron where to click and what to expect, adding a line to the log file, that indicates a different type of use, as well as when people follow a link in error. Finally, researchers who use a particular resource on a regular basis are likely to have that file bookmarked, therefore bypassing our logging process altogether and going directly to the resource. Despite all the places where errors can be introduced into the data, there are many positives to the numbers being gathered. The data is consistent enough for meaningful comparison, and include users information not available from vendors.

Libraries are likely facing an upcoming budget crunch for support of electronic products similar to that experienced for journal subscriptions in the early to mid-1990s. Most librarians, whether or not directly involved with collection development, are aware of libraries experiencing journal reductions in the past. As budgets become tight again, many institutions are facing these issues, only this time electronic resources are also being scrutinized. Before a money problem arises, data is needed to understand what resources are being used, and perhaps learn why some of the important files are underutilized. The numbers can help libraries initiate conversations with users about their needs and where they are being met, and also help us know what files should be marketed and to which audiences. As with any collection development decision, the numbers will help support a choice, but would never be the sole factor in deciding resource reductions. Research, curricular and institutional needs are among the criteria for evaluation and ultimate decision.

The study began with a request from the Management and Economics Library at Purdue University to gather click-through numbers on their electronic abstract and index resources. Initial analysis was done with a free program called *Analog* that analyzes Web log files on a PC running almost any operating system. The software was chosen for its ability to run on a Windows NT machine and the opportunity for the analyst to specify the format of the log file to be analyzed. The output from Analog is an HTML file that shows the most requested URLs, ranked highest to lowest, and how many times during a given period they were requested. The time period presented is determined by the dates included in the log file. For example, if the log file is divided into one-month segments, then each analysis provides information for use during that month. Other information included the heaviest use times of day and days of the week. This analysis provided the primary information desired.

After working with the information for the Management Library, the Engineering Library staff recognized the applicability and asked to be included in the tracking. A short addition–the CGI script call mentioned earlier–was made to the Engineering Libraries links for those resources to be tracked. Adding a piece to each URL is not as overwhelming as it may sound, provided the library uses a Web-site management program. In such programs, it is possible to change a link in one location and have all occurrences of that URL change throughout the site. The log file analysis, now being done by the Libraries' Information Technology Department rather than the Engineering Library, uses a package called *WebTrends®*. The software can output the analysis information in an Excel spreadsheet, allowing additional manipulation of the information. Another advantage of *WebTrends®* is the ability to designate the specificity of analysis on the originating IP addresses for each request.

Figure 1 shows the brief perl script that creates the log file and redirects the user to the requested resource. Analysis of the log file generates the statistics. Since the nature of the logging is minimal, transferring a very small amount of information each time it writes to the log, the user does not see any lag time before connecting to the resource to be used.

Figure 2 includes sample lines from the log file. The sample shows the information tracked, which includes the IP address of the requesting machine, the originating URL, the date and time of the request, the URL that is being requested, and the protocol being used. The last numbers represent the status of the request and the number of bytes sent in response to the request (Dowling 2001). The client information, which includes IP address and referring page, is freely available information gathered from the Web browser as part of the HTTP protocol.

The first run of data included April through mid-September 2001 and provided general numbers which presented largely what was expected. Reviewing the report generated by *WebTrends®* showed that our most

FIGURE 2. Sample Lines from the Log File

```
168.229.4.1 - http://thorplus.lib.purdue.edu/engr/civil.html -
[01/Oct/2001:10:11:33 -0500] "GET http://hwwilsonweb.com/ HTTP/1.0" 301 1
128.210.124.40 - http://www.lib.purdue.edu/engr/trindexes.html -
[01/Oct/2001:10:37:27 -0500] "GET http://212.49.195.109/webCD/CGI.EXE
HTTP/1.0" 301 1
```

requested resources were those that provide the full-text of vendor catalogs and standards from IHS. A study done in 1986 showed that practicing engineers used product catalogs as their primary information resource (Jones 1986). At this time the URL for the full-text standards and the product catalogs is the same, so future work will include separating these two resources to determine which is creating the highest use. Looking at the vendor-supplied information on the number of logins indicates that the majority of the use for these resources is being generated by the online standards, nearly three times the product catalog use for this time period. It would be interesting to see the 1986 study redone in light of more materials being available online. Second, by order of most requested items, are the primary engineering databases, Compendex® (Engineering Index online) and INSPEC®. This finding is also in line with the 1986 study by Jones and LeBold. The uses of these files are being logged separately through unique URLs, but *WebTrends®* is truncating the URL requested before the unique part of the string is read. As a result, it is not possible to determine if Compendex® or INSPEC® is used more often. Further refining of the *WebTrends®* profile should address this problem.

Why is this data gathering of interest now? Constant budget questions drive a librarian's desire to quantify where libraries get the "biggest bang for our buck" and which databases could potentially be dropped if the money was not available. The need for comparative statistics, which cannot be determined from vendor supplied data at this point, created the need to develop a process of our own. The use of a redirect log to count click-throughs fills the data need at a basic level and provides user information unavailable elsewhere. This data gathering provides a good starting place for learning which of the resources the library provides are being used and how often that use occurs.

# REFERENCES

Burford, Nancy and Heather Goetz. 2001. Tracking E-Journal Use from the OPAC. *Poster session presented at the Voyager User Group Meeting*, Des Plaines, IL, April 20-21.

Covey, Denise Troll. 2002. Usage Studies of Electronic Resources. Chap. 3 in *Usage and Usability Assessment: Library Practices and Concerns.* Washington, DC: Digital Library Federation, Council on Library and Information Resources.

Dowling, Thomas. 2001. Lies, Damned Lies and Web Logs. *School Library Journal: netConnect* (Spring):34-35.

International Coalition of Library Consortia. 1998. *Guidelines for statistical measures of usage of web-based indexed, abstracted, and full-text resources.* http://www.library.yale.edu/consortia/webstats.html.

Jones, R., and W.K. Lebold. 1986. Keeping Up to Date and Solving Problems in Engineering. In *Proceedings, 1986 World Conference on Continuing Engineering Education.* Lake Buena Vista, FL.

Wonsik "Jeff" Shim, Charles R. McClure, and John Carlo Bertot. 2000. *ARL E-Metrics Project: Developing Statistics and Performance Measures to Describe Electronic Information Services and Resources for ARL Libraries: Phase One Report.* Tallahassee, FL: Information Use Management and Policy Institute, School of Information Studies, Florida State University. http://www.arl.org/stats/newmeas/emetrics/phaseone.pdf.

Wonsik "Jeff" Shim et al. 2001. *Measures and Statistics for Research Library Networked Services: Procedures and Issues. ARL E-metrics: Phase II Report.* Tallahassee, FL: Information Use Management and Policy Institute, School of Information Studies, Florida State University. http://www.arl.org/stats/newmeas/emetrics/phasetwo.pdf.

# Capturing Patron Selections from an Engineering Library Public Terminal Menu: An Analysis of Results

William H. Mischo
Mary C. Schlembach

**SUMMARY.** The Grainger Engineering Library at the University of Illinois at Urbana-Champaign has developed a mechanism to capture and record user resource selection from the Grainger public terminal menu using a 'click-through' redirect transaction log technique. The transaction log record includes the Internet Protocol (IP) address of the source computer and an indication of the resource selected, whether it is located on a local or remote server. While there are limitations in this approach, this data is useful for collection assessment, user service needs analysis, and interface design purposes. An analysis of 84,824 public terminal menu selections over the period from January 2001 to May 2002 was performed. During this period, 67.8% of the recorded selections came

William H. Mischo is Head, Grainger Engineering Library Information Center and Professor of Library Administration, University of Illinois at Urbana-Champaign, Urbana, IL (E-mail: w-mischo@uiuc.edu). Mary C. Schlembach is Assistant Engineering Librarian, Grainger Engineering Library Information Center and Associate Professor of Library Administration, University of Illinois, Urbana-Champaign, Urbana, IL (E-mail: schlemba@uiuc.edu).

[Haworth co-indexing entry note]: "Capturing Patron Selections from an Engineering Library Public Terminal Menu: An Analysis of Results." Mischo, William H., and Mary C. Schlembach. Co-published simultaneously in *Science & Technology Libraries* (The Haworth Information Press, an imprint of The Haworth Press, Inc.) Vol. 21, No. 1/2, 2001, pp. 127-138; and: *Information Practice in Science and Technology: Evolving Challenges and New Directions* (ed: Mary C. Schlembach) The Haworth Information Press, an imprint of The Haworth Press, Inc., 2001, pp. 127-138. Single or multiple copies of this article are available for a fee from The Haworth Document Delivery Service [1-800-HAWORTH, 9:00 a.m. - 5:00 p.m. (EST). E-mail address: docdelivery@haworthpress.com].

http://www.haworthpress.com/store/product.asp?sku=J122
10.1300/J122v21n01_11

from Grainger in-library computers, indicating heavy use of the public terminal page by remote users. Online catalog usage (at 35.6% of the total selections) and Abstracting and Indexing (A & I) Services usage (at an aggregate 34.6%) were high. Periodical access selections represented, at the minimum, 47.6% of the total selections and direct electronic journal links were 14.9% of the total. Patrons are making extensive use of publisher full-text repository sites. *[Article copies available for a fee from The Haworth Document Delivery Service: 1-800-HAWORTH. E-mail address: <docdelivery@haworthpress.com> Website: <http://www.HaworthPress.com> © 2001 by The Haworth Press, Inc. All rights reserved.]*

**KEYWORDS.** Transaction logs, public terminal menus, Grainger Engineering Library, end-user searching behavior

## INTRODUCTION

Library public terminals provide users with a rich portal function that provides access to myriad remote and local information resources. These resources typically include local, state, and regional Online Public Access Catalogs (OPACs), national shared cataloging databases such as Online Computer Library Center (OCLC), remote and locally loaded Abstracting and Indexing (A & I) services, licensed publisher full-text repositories and portals, and custom local databases and Web sites produced by the library. Clearly, it is important for libraries to gather usage statistics on these resources for collection development decisions and to help identify public service needs. Determining the frequency and order of selections from the public terminal menu can provide useful interface design data for the arrangement of menu options and the determination of primary and secondary menus. In particular, libraries spend a significant portion of their collection development budget on a variety of licensed A & I services and a growing number of full-text publisher sites. These resources may be paid out of the budgets of branch or departmental science and engineering libraries or from central library funds, or a combination of both. Determining usage data for these licensed resources is of primary importance.

To determine online resource usage, libraries have historically relied on OPAC transaction logs and vendor and publisher supplied usage statistics. There is a rich literature on library transaction log analysis (Peters 1996, Tarr 2001). With vendor supplied usage data, there have been significant problems with incompleteness, timeliness, and inconsistent format of the vendor supplied usage data. However, this is presently the most reliable measure of the use of licensed Web-based online resources (Blecic, Fiscella, and Wiberley 2001).

As a means of determining specific resource selection and usage at the point of contact with patrons, the Grainger Engineering Library at the University of Illinois at Urbana-Champaign (UIUC) has developed a mechanism to record user resource selection from the Grainger public terminal menu. This technique complements the data provided by vendors and provides some on-site evidence-based data for determining resource usage.

The UIUC Library provides custom public terminal Web page menus at each of the forty subject departmental libraries. Each departmental library constructs Web-page menus that emphasize the information resources that they feel best meet the information needs of their specific clientele. The Grainger Library has 22 public terminals on five floors, with a concentration in the Reference desk area. In addition, the Grainger Library has 30 PCs in a microcomputer cluster and another 62 UNIX workstations in two engineering workstation labs. The Grainger Library also provides walkup Internet access at 32 carrels and group study rooms for users with laptop computers.

The Grainger Library has been recording statistics on user resource selection from the public terminal menu since late 1997. This paper will report on and analyze selections made during a seventeen-month period from January 2001 through May 2002 from the public terminal Web menu. These menu selections have been made from public terminals and other PCs in the Grainger Library and from remote computers accessing the public terminal Web page.

## METHODOLOGY

Grainger Library public terminal Web page menu selections are recorded by using a 'click-through' transaction recording technique. Every time a patron clicks on a specific resource on the page, a transaction log entry is written to a file on a Grainger Library Web server. The transaction entries are written to separate files based on the IP address of the public terminal or user workstation from which the selection is being made. Because the IP address of the browser machine is being recorded, we are able to determine if the connection is being made from an in-library computer or from a remote workstation using the public terminal menu. In addition, all selections made from each public terminal are stored in separate, discrete files. This provides a convenient means to compare public terminal usage in different areas of the Grainger Library and to differentiate individual transactions on a specific public terminal or workstation to perform a time series analysis of usage.

It is important to note that Web server logs only record usages of pages that are stored on that specific Web server. The typical library Web page–such as the public terminal page discussed here–contains links not only to pages that are on the same server or other local servers but also to pages that are elsewhere on the Web, such as OPACs, A & I services, and publisher sites. Clicking on one of the links to an off-site page will not generate a log entry on the local Web server that contains the public terminal Web page. Typically, only the logs of the remote server will record these transactions. Web server transaction log issues are discussed at length in another article in this volume (Tarr 2001).

The menu selection recording technique employed here uses a redirect method that sends each selection through a local Web server log page containing server-side scripting code. This log page invokes a dynamic link library (.dll) running on the Web server to record the resource selected from the extended URL and then redirects the browser to the appropriate resource URL. An alternate scripting redirect method employed at Purdue University is presented in this volume (Van Epps 2002).

On the Grainger public terminal Web page shown in Figure 1, the URL behind the *INSPEC Database (1969-)* menu selection is in the form:

*http://shiva.grainger.uiuc.edu/top/log.asp?insp-http://ovid.grainger.uiuc.edu/Inspec* . . .

All public terminal selections are first sent to the *log.asp* page with a querystring containing a four letter code for the resource selected (in this case *insp*) and the URL for the selected resource. The scripting code on the *log.asp* page invokes the .dll program that writes the transaction information to a file on the same server and then redirects the browser to the http://ovid.grainger.uiuc.edu/Inspec . . . URL. The transaction is recorded and the redirect takes place instantaneously and in a manner that is totally transparent to the user.

The transaction log entry contains the date and time of the transaction, the specific IP address, and the name of the resource that has been selected. A sample one-line transaction recording the selection of the INSPEC A & I service appears as follows:

*10-17-2001  21:18:28  128.174.36.184  INSPEC.*

FIGURE 1. Grainger Public Terminal Menu

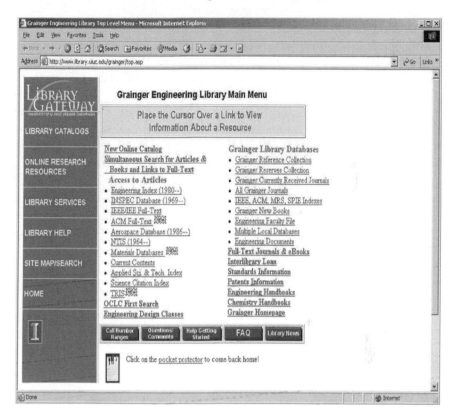

The IP address from which this transaction occurred is *128.174.36.184* and the *INSPEC* selection is derived from the *insp* four letter code from above. This transaction line was written into a file named *tranl36184*. These files are collected at regular intervals and for analysis purposes can be merged into one large file. A Visual Basic program then reads through the file and adds together and stores the number of usages for each resource. Because the resource selection transaction logs record the IP address of the public terminal menu user, it can be determined if the transaction is taking place from a Grainger Library public terminal or from a user accessing the public terminal menu from a remote site, such as a laboratory, office, home, or outside computer lab. Custom Visual Basic programs perform more detailed analyses by breaking usage down by each public terminal and comparing usage by public terminal vs. remote workstations.

There are some important limitations with determining overall resource usage using this menu redirect approach. The recorded data only tells us which link a user clicked on from the public terminal menu at a specific point in time. It does not provide us with any navigation or click-through information after a patron has reached the specific target site or any click-through data from any other subsequent local or remote page. There is no accurate way to determine how long a patron was using a resource, other than by looking at the time stamps to determine when another selection was made from the main public terminal menu. And of course, this method does not record any resource usage by patrons who have bookmarked a specific menu item such as Compendex or IEEE Electronic Library (IEL) on their office or home workstations. The public terminals are re-booted every fifteen minutes so persistent bookmarks cannot be established on the public terminals.

Grainger Library reference desk terminals use a different resource menu and selections made from the reference desk menu are not recorded. In addition, Grainger staff machines typically point to the main home page as the default page. However, a selection made by a reference staff member assisting a patron at a public terminal is recorded and cannot be distinguished from a patron selection. The Grainger Library main home page contains a prominently displayed link to the public terminal page. Patrons who are being assisted at the public terminals are directed to use the public terminal page or the link to it from the Grainger Home Page to facilitate their remote work. In this way, we are attempting to isolate the public terminal menu as a jumping-off point for patron library research.

Also we are not presently recording any URLs entered by the patron on the Web browser. A study to more completely examine patron public terminal behavior by recording both menu selections and patron-entered URLs is just beginning. The method of recording patron resource selections reported here provides us with a homogeneous, consistent mechanism for determining online resource selection from the public terminal Web page menu.

An alternate method for determining public terminal usage and public terminal menu selections is to observe specific public terminal activity at fixed time intervals and record the specific URL that the user is visiting at that point in time (Konomos and Herrington 2000).

## RESOURCE USAGE

The Grainger Engineering Library provides primary research and instructional support for the eleven UIUC College of Engineering departments (except for the Department of Physics), which includes computer science, but excludes chemical engineering and agricultural engineering. The Grainger Library directly supports nearly 5,300 undergraduate engineering students and 1,950 graduate students. The current Grainger public terminal menu is shown in Figure 1.

Using the methods described above, a total of 84,824 transactions or clicks were recorded from January 2001 through June of 2002 of 34 of the menu items shown above. The *Chemical Abstracts Sci-Finder* client is only located on one public terminal, but is included in the study. The *EngnetBase, ChemnetBase,* and *Simultaneous Search* options were not available for the entire period of the study and are not included in the analysis. However, the *Simultaneous Search* system, which provides the capability of searching four A & I Service databases, the OPAC, the Grainger New Books database, and the Google Web search engine asynchronously was selected 360 times in a month and a half of availability.

The selected menu items, in order of frequency, are shown in Table 1.

## ANALYSIS OF RESULTS

During the time period covered by the study, 67.8% of the public terminal menu selections were made from public terminals or computers within the Grainger Library, including PC cluster machines, user walkup laptops, or Grainger engineering workstations. The percentage of in-library selections has gone down in the last several years; for example, including the year 2000 in the study raises the overall in-library selection percentage to 73.6%. The fact that 32.2% of the selections were made from machines outside the Grainger Library indicates a growing number of users accessing the public terminal menu from laboratories, dorms, offices, homes, and other remote sites. Factoring in all users with bookmarked pages to public menu selections or users accessing resources through the main home page only serves to illustrate the large number of remote users of library resources.

The OPAC was the most frequently selected single resource. However, at 35.6% of the total items selected, OPAC usage represents significantly less than half of the menu items chosen. A number of

## TABLE 1. Menu Item Selections

| Resource | Frequency | Percentage of Total |
|---|---|---|
| Online Catalog | 30,226 | 35.6% |
| Compendex | 7,766 | 9.2% |
| IEL (IEEE/IEE) | 7,399 | 8.7% |
| All Grainger Journals Ever Received Database | 4,650 | 5.5% |
| INSPEC | 3,895 | 4.6% |
| Currently Received Grainger Journals Check-In Database | 3,501 | 4.1% |
| Electronic Resources Page | 3,472 | 4.1% |
| Web of Science | 2,423 | 2.9% |
| Grainger Reference Collection Index | 1,846 | 2.2% |
| OCLC First Search | 1,803 | 2.1% |
| ACM Portal | 1,745 | 2.1% |
| Interlibrary Loan | 1,551 | 1.8% |
| Applied Science & Technology Index | 1,424 | 1.7% |
| Materials Database | 1,402 | 1.6% |
| Current Contents | 1,194 | 1.4% |
| Grainger Main Home Page | 1,135 | 1.3% |
| Aerospace Database | 950 | 1.1% |
| Grainger Reserves Database | 894 | 1.1% |
| UIUC Engineering Documents Collection | 828 | 1.0% |
| Design Classes Site | 793 | 0.9% |
| Help Getting Started Page | 731 | 0.9% |
| NTIS Database | 589 | 0.7% |
| Standards Site | 565 | 0.7% |
| Grainger New Books Database | 528 | 0.6% |
| Patents Site | 522 | 0.6% |
| Society Publications Index | 519 | 0.6% |
| TRIS Database | 491 | 0.6% |
| UIUC Engineering Faculty Interest File | 473 | 0.6% |
| Call Numbers (Grainger Shelving Chart) | 385 | 0.5% |
| Frequently Asked Questions Database | 333 | 0.4% |
| Multiple Local Databases Search | 287 | 0.3% |
| Library News | 241 | 0.3% |
| Make a Comment | 213 | 0.3% |
| Chemical Abstracts | 50 | 0.1% |

transaction log studies in the early 1990s of extended online catalogs that incorporated locally mounted periodical index databases showed a reduced percentage of OPAC searching (Potter 1989; Mischo and Cole 1992).

Several of the other menu resources complement the OPAC selections by also indicating patron attempts to access the monographic literature. These include the OCLC First Search database, the Grainger New Books database, and the Grainger reference collection database. While a (probably large) percentage of the OPAC and First Search users came in looking to locate a serial title, adding these additional usages to the OPAC selections brings the upper-end aggregate monographic-related usages to 34,403 or 40.5% of the total selections.

A large percentage of the selections clearly represent user attempts to access the periodical literature in order to retrieve articles on a topic or by an author, to locate library holdings and availability of a specific journal, or to determine if an online full-text version of a journal or specific article is available. A number of studies of the information seeking behavior of scientists and engineers show increased usage of the journal literature and, particularly, of the electronic journal literature (Tenopir and King 2001; Hiller 2002).

In this study, the transaction logs show that several of the A & I Services and periodical index databases were frequently selected, particularly Compendex, IEL, and INSPEC. Taken together, the A & I services and periodical portals represented by Compendex, IEL, INSPEC, Web of Science, Materials Database, Applied Science and Technology Index (ASTI), Current Contents, Aerospace Database, the Association for Computing Machinery (ACM) Portal, National Technical Information Service (NTIS), Transportation Research Information Services (TRIS), and Chemical Abstracts totaled 29,328 selections or 34.6% of the total menu selections. When all the periodical menu resources are considered, including the resources listed above (minus NTIS) plus the All Grainger Journals, the Currently Received Grainger Journals, and the E-resources database registry, the total usages increase to 40,362 selections or 47.6% of the total. As mentioned above, many of the OPAC selections were made with the intention of locating a periodical resource. If half of the OPAC selections were added to the periodical selection group, this would bring the aggregate periodical use selections to 55,362 or 65.2% of the choices. In analyzing periodical related selections, it is unclear what roles the limitations and feature set of the UIUC Library OPAC play. At the time of this study, the UIUC Library employed the Data Research Associates (DRA) Classic online catalog without the serials check-in module and with very few OPAC e-journal links included.

The last several years have witnessed a significant growth of interest in electronic journals. From the Grainger public terminal menu, direct e-journal access is provided by the Electronic Resources list, the IEL site, and the ACM Portal. Together, these resources were chosen 12,616 times representing 14.9% of the total selections. The high use of the IEEE IEL page and the ACM Portal illustrates the popularity of publisher-based full-text search and discovery sites. One could argue that we should be looking at presenting top-level menu access to other publisher repositories, such as ScienceDirect.

In addition, the Grainger Currently Received Journals database also contains links to available electronic full-text versions of journals. Adding the Currently Received Journals selections to this e-journal group raises the total to 16,117 selections or 19% of the total. There is some evidence that users prefer to access e-journals from a separate registry rather than through the OPAC (Hurd 2001). Again the limitations in access mechanisms for e-journals present in the UIUC Library OPAC play an obvious role here.

The Grainger Library public terminal menu contains links to a number of Web-based locally produced databases that complement the OPAC and other information resources, both online and print. These local databases are described in detail in a previous article (Mischo and Schlembach 1999). The local databases, comprised of the All Journals database, the Currently Received Journals database, the E-resources registry, the Reference Collection index, the Reserves database, the College of Engineering Documents Collection, the Society Publications index, the New Books database, the Engineering Faculty Interest file, Frequently Asked Questions (FAQ) database, and the combined Multiple Local Databases search system were used 18,885 times making up 22.2% of the total usage.

Table 2 provides a summary of the menu selections, grouped into general categories of usage. Some of these resource selections fit into more than one category.

## CONCLUSION

Grainger Library public terminal Web page menu selections are being captured by using a click-through redirect transaction recording technique that sends each selection through a local Web server log page.

TABLE 2. Selections Arranged by General Category

| Category | Total Uses | Percentage of Total |
|---|---|---|
| OPAC, Monographic Database Usage (Upper Limit) | 34,403 | 40.5% |
| A & I Services Usage | 29,328 | 34.6% |
| Access to Periodical Info | 40,362-55,362 | 47.6%-65.2% |
| E-Journals Access | 16,117 | 19% |
| Local Databases Usage | 18,885 | 22.2% |

This technique provides a mechanism to easily record the selection of resources on both local and remote servers. This technique also provides a means to record the IP address of the computer using the menu page. During a seventeen-month period covering 2001 and early 2002, 67.8% of the recorded selections came from Grainger in-library computers. The large number of remote selections (32.2%) indicates heavy use of the public terminal page by remote users.

An analysis of the recorded selections from the public terminal Web page shows that 35.6% of the total menu selections were of the Online Catalog, the highest for any individual menu item. However, periodical related usage was also very high. Of the selections made, 34.6% were to A & I service databases and publisher periodical portals. When all periodical access resources are considered, this percentage increased to, at minimum, 47.6%.

User selection of links related to electronic journals was also very high. The selection of links providing direct access to e-resources represented 14.9% of the total. The use of publisher full-text repositories, in the form of IEL and the ACM Portal, was very high. When including links to the Currently Received Journals database, the percentage of e-journal related selections increased to 19%. It is clear that users are very interested in access to available electronic full-text versions of journals.

In addition, the Grainger Library locally generated databases were heavily utilized. These local databases, which provide access to custom bibliographic and information resources, represented 22.2% of the total selections.

It is clear that the Web page menu selection recording technique reported here provides useful information on user searching behavior. These usage statistics provide important information for collection development decisions and help to identify public service needs. In addition, the data can also provide useful interface design data that can be used in the arrangement of menu options and the determination of primary and secondary menus.

The large number of remote users accessing the public terminal page presents a challenge to the library in terms of instruction on database selection, search and navigation assistance, and remote instructional activities.

## REFERENCES

Blecic, Deborah C. and Joan B. Fiscella, and Stephen Wiberley. 2001. The Measurement of Use of Web-Based Information Resources: An Early Look at Vendor-Supplied Data. *College and Research Libraries* 62(5):434-453.

Hiller, Steve. 2002. How Different Are They? A Comparison by Academic Area of Library Use, Priorities, and Information Needs at the University of Washington. *Issues in Science and Technology Librarianship* Winter 2002 [cited 12 October 2002] Available from http://www.istl.org/istl/02-winter/article1.html.

Hurd, Julie M. 2001. Digital Collections: Acceptance and Use in a Research Community. Crossing the Divide: *Proceedings of the Tenth National Conference of the Association of College and Research Libraries*, ed. Hugh A. Thompson. Chicago: ACRL, 312-319.

Konomos, Philip and Scott Herrington. 2000. Evaluating the Use of Public PC Workstations at the Arizona State University Libraries. *The Electronic Library* 18(6): 403-406.

Mischo, William H. and Mary C. Schlembach. 1999. Web-based Access to Locally Developed Databases. *Library Computing* 19(1):51-58.

Mischo, William H. and Timothy W. Cole. 1992. The Illinois Extended OPAC: Library Information Workstation Design and Development. *Advances in Online Public Access Catalogs, Volume 1*, ed. Marsha Ra. Westport: Meckler, 38-57.

Peters, Tom. 1996. Using Transaction Log Analysis for Library Management Information. *Library Administration and Management* 10(1):20-25.

Potter, William Gray. 1989. Expanding the Online Catalog. *Information Technology and Libraries* 8(2):99-104.

Tarr, Beth L. 2001. Looking for Numbers with Meaning: Using Server Logs to Generate Web Site Usage Statistics at the University of Illinois Chemistry Library. *Science & Technology Libraries* 21(1/2): 139-152.

Tenopir, Carol and Donald W. King. 2001. Electronic Journals: How User Behaviour is Changing. *Online Information 2001. Proceedings of the International Online Information Meeting*. Oxford: Learned Information Europe, 175-181.

Van Epps, Amy S. 2001. When Vendor Statistics Are Not Enough: Determining Use of Electronic Databases. *Science & Technology Libraries* 21(1/2): 119-126.

# Looking for Numbers with Meaning:
# Using Server Logs
# to Generate Web Site Usage Statistics
# at the University of Illinois
# Chemistry Library

Beth L. Tarr

**SUMMARY.** Since the introduction of the first computerized information retrieval systems in libraries, librarians have attempted to gather statistical information on system usage. For Web sites, the most important source of this information is the Web server transaction log. This paper explores the reasons behind the search for Web usage statistics, the limitations of Web server logs and the methods used to analyze them, and

Beth L. Tarr, MS (Library and Information Science), BA (Liberal Arts and Sciences), is librarian, Los Angeles Public Library (E-mail: btarr@lapl.org). While completing her MS, she was a graduate assistant at the Unversity of Illinois Chemistry Library.

The author would like to thank Tina Chrzastowski, Chemistry Librarian and Professor of Library Administration, and David Dubin, Senior Research Scientist, Information Systems Research Laboratory, both of the University of Illinois at Urbana-Champaign, for their guidance in researching this paper.

[Haworth co-indexing entry note]: "Looking for Numbers with Meaning: Using Server Logs to Generate Web Site Usage Statistics at the University of Illinois Chemistry Library." Tarr, Beth L. Co-published simultaneously in *Science & Technology Libraries* (The Haworth Information Press, an imprint of The Haworth Press, Inc.) Vol. 21, No. 1/2, 2001, pp. 139-152; and: *Information Practice in Science and Technology: Evolving Challenges and New Directions* (ed: Mary C. Schlembach) The Haworth Information Press, an imprint of The Haworth Press, Inc., 2001, pp. 139-152. Single or multiple copies of this article are available for a fee from The Haworth Document Delivery Service [1-800-HAWORTH, 9:00 a.m. - 5:00 p.m. (EST). E-mail address: docdelivery@haworthpress.com].

possible ways to gather more meaningful Web usage statistics, taking into account the particular challenges Web servers present for electronic reserves. *[Article copies available for a fee from The Haworth Document Delivery Service: 1-800-HAWORTH. E-mail address: <docdelivery@haworthpress. com> Website: <http://www.HaworthPress.com> © 2001 by The Haworth Press, Inc. All rights reserved.]*

**KEYWORDS.** Electronic reserves, academic library, transaction logs, transaction analysis, usage statistics, library Web site

## QUANTIFYING USAGE:
## THE SEARCH FOR THE PERFECT NUMBERS

Quantifying usage of materials and services is a familiar topic for libraries. Many libraries count the number of patrons who enter the library in a day, the number of reference questions asked in a week, and the number of books checked out in a year. At the University of Illinois at Urbana-Champaign (UIUC) Chemistry Library, manual daily statistics are kept on the numbers of directional and reference questions asked by patrons at the desk or by telephone; class reserve folders checked out; and bound journals, unbound journals, circulating books, and reference books reshelved. Statistics are also kept on the numbers of photocopies made in the library and books checked out through Interlibrary Loan.

This experience collecting usage statistics led to an expectation that when the Chemistry Library starting putting electronic reserve materials (full-text scanned PDF files) online through its Web site, it would be able to quantify the use of those materials. Since the library system Web server maintains a transaction log, it seemed like a simple matter of retrieving the number of accesses from the server log. Like many libraries, though, the Chemistry Library quickly discovered that acquiring meaningful statistics on the use of materials available via the World Wide Web is not an easy task.

This paper explores the reasons behind the need for Web usage statistics, the limitations of Web server logs and the methods used to analyze Web server logs, and possible ways to gather more meaningful Web usage statistics, taking into account the particular challenges Web servers present.

## WHY TRACK WEB USAGE STATISTICS?

The first question to ask in developing methods for gathering statistics on Web site usage is the most basic question: Why gather statistics on Web site usage at all? Why bother monitoring how many people use a particular Web site or how those people use it? As the difficulties in obtaining reliable and meaningful statistics become apparent, these questions become more pressing.

The reasons for wanting statistics about Web site usage vary, depending largely on the motivation behind the Web site itself. Mobasher et al. noted that, for commercial Web sites, "analysis of server access data can provide information on how to restructure a Web site for increased effectiveness, better management of workgroup communication, and analyzing user access patterns to target ads to specific groups of users" (Mobasher, Jain, Han & Srivastava 1996). For libraries, the two most significant reasons for keeping track of Web site usage are accountability and improving service (Johnson 2000).

When a library makes information available on the Web, it wants to ensure that the information is useful and is being used. Resources on the Web are part of that library's collection, and library staff invest time and resources in creating and maintaining the Web site. If the information is not being used, it is an unnecessary drain on library resources. Ideally, the library's Web site will increase the library's user base, and usage statistics provide quantifiable data about that user base.

The ability to monitor usage of the Web site can also help the library staff determine needed developments and improvements. Popular pages can be highlighted on the library's home page for new users. Pages that receive unexpectedly low usage can be spotted and redesigned or removed. Maintenance of the Web site should be part of the library's collection development plan, or the library risks failing to meet the needs of remote users (Johnson 2000).

For the UIUC Chemistry Library, monitoring the use of the library Web site took on new importance with the introduction of the electronic reserves. The electronic reserves consist of a variety of materials, primarily HTML and PDF documents, but also including static image and animation files, kept as reserve materials for classes in Biochemistry, Chemical Engineering, and Chemistry. At the beginning of each semester, professors may request electronic reserves for their classes. An index page is created for each class, featuring hypertext links to the class reserve materials. Access to the electronic reserves is limited to users within the University of Illinois' range of IP addresses; off-campus users must log into a proxy server using a University ID and password. Some classes have both print reserves (including books, articles, and quiz and homework solutions) and electronic reserves, some classes have only print reserves, and some classes have only electronic reserves. In order to remain consistent with the print reserve usage statistics kept both manually and through the online catalog system, the Chemistry Library staff wanted to obtain statistics on electronic reserves usage. This required a new method of gathering statistics: transaction log analysis (TLA).

## A LOOK BACK AT THE HISTORY
## OF TRANSACTION LOG ANALYSIS IN LIBRARIES

For the past several decades, ever since the first computerized information retrieval (IR) systems appeared in libraries, librarians have studied how these automated systems are used (Cochrane & Markey 1983; Simpson 1989). Many of these studies have analyzed transaction logs as an unobtrusive way of monitoring both the performance of the system itself and the way it was used by human information seekers (Peters 1993a). Of these two goals, monitoring the human behavior has proven to be the more difficult task, as the transaction logs being studied were intended as records of system performance, not human behavior.

Peters reviewed several areas of transaction log analysis research including studies of the frequency of use of particular commands, the differences between vocabulary used by information seekers and vocabulary used to classify system records, the response times of the IR systems and of the human users, and user persistence when faced with uncooperative systems. Peters identified two problems in using transaction log analysis: "the difficulty in establishing the parameters of a search session and the inability to tie the observed searching behavior with the needs, thoughts, opinions, goals, emotions, and evaluations of the users engaged in the observed searching behavior" (Peters 1993a). Kurth also noted that data in transaction logs "effectively describe what searches patrons enter and when they enter them, but they don't reflect, except through inference, who enters the searches, why they enter them, and how satisfied they are with their results" (Kurth 1993). The studies could tell researchers how people were using the system, but they could not tell researchers how satisfied users were with the system. As Kaske concludes, the quantitative data from system logs must be combined with qualitative data obtained through other means, such as interviews, visual observation, and questionnaires, to create a complete picture of system usage (Kaske 1993).

Increasing use of the World Wide Web has not eliminated these problems. In fact, it has made it more difficult for libraries to analyze the use of and user satisfaction with their electronic resources. Instead of coming to the library, where they can be observed using the system, patrons can access the resources from home, offices, laboratories, and dormitories. Flaherty noted that it was difficult to identify the use of the system by individual patrons through the data in the transaction logs because "it is usually not possible to know with certainty when a patron's activity begins and ends" (Flaherty 1993). The use of HyperText Transfer Protocol (HTTP), the data transmission format on which the World Wide Web relies, has made this task nearly impossible. HTTP is a stateless protocol, meaning that each request a user makes to the server is treated as a separate transaction, and individual users (who are not required to log into or out of the system) are almost impossible to identify. Peters et al. reported that server logs in past library TLA studies consisted of "electronically recorded interactions between online information retrieval systems and the persons who search for the information found in those systems," now record the interactions between one computer and another computer (Peters, Kurth, Flaherty, Sandore & Kaske 1993b).

## *INSIDE THE LOG:*
## *WHAT SERVER LOGS RECORD*

Most Web server logs follow the HTTP Common Log Format, which records the IP address of the client computer, the name of the user (if available), the user password (if available), the date and time of the request, the method and URL requested, the protocol (and version) specified by the client, an error or status code, and the size (in bytes) of the file transferred by the server. One line in a log file using Common Log Format, then, would look like this:

```
130.126.33.1 - - [04/29/2001:12:08:27] "GET /chx/reserves/chem331/solutions34.pdf"
HTTP/1.1 200 95967
```

This format is not followed by all server logs. Microsoft's Internet Information Server (IIS) uses the World Wide Web Consortium (W3C) extended log format. An entry in an IIS 4.0 Web server log might look like this example from IBM:

```
#Software: Microsoft Internet Information Server 4.0
#Version: 1.0
#Date: 1998-11-19 22:48:39
#Fields: date time c-ip cs-username s-ip cs-method cs-uri-stem cs-uri-query sc-status sc-
bytes cs-bytes time-taken cs-version cs(User-Agent) cs(Cookie) cs(Referrer)

1998-11-19 22:48:39 206.175.82.5 - 208.201.133.173 GET /global/images/navlineboards.gif -
200 540 324 157 HTTP/1.0
Mozilla/4.0+(compatible;+MSIE+4.01;+Windows+95) USERID=CustomerA;+IMPID=01234
http://yourturn.rollingstone.com/webx?98@@webx1.html
```

Additionally, not all Web server logs are kept as text files. IIS 4.0 also offers the option of keeping server logs in a database. These log entries use Open Database Connectivity (ODBC) format and are saved directly to a database file.

The University of Illinois Library system uses this IIS-ODBC format and saves its Web server access logs as tables in a Microsoft SQL server database. Three tables are kept for the year: Summer, Fall, and Spring terms; and access to the data is restricted to Library staff. Earlier data is preserved in tape storage and is not readily available for access. To retrieve data from tape storage, Library staff must contact the Library Systems Office. Library Systems staff are able to query the active database (the most recent three academic terms) directly, but other users must access the log data through an Active Server Page (ASP) query form. The form offers some options for creating specific queries, but users are still somewhat limited (Lewenberg 2002).

The search form does not retrieve all possible fields. An entry in the database, with the client IP address resolved, looks like this:

*Request 1 of 5478:* /chx/reserves/chem331/solutions34.pdf (GET)
    by Client: www-s2.library.uiuc.edu at: 4/29/2001 12:08:27 AM Status: 200
    Bytes: 95967

## MISSING INFORMATION:
## WHAT SERVER LOGS DO NOT RECORD

Web server logs are a rich source of data about how the Web server is being used. Web server usage, however, is not necessarily equal to the usage of the resources (Web pages) stored by that server. As with the earlier information retrieval system analyses, it is difficult, if not impossible, to track individual users and follow their interactions with the Web server. Users with dial-up accounts and cable modems are sometimes assigned a temporary IP address by their Internet Service Provider (ISP). This IP address lasts for the duration of the connection; the next time the user connects, he is assigned a different IP address. At the other extreme, many users may access the site through a firewall or proxy server; all of these users would appear to have the same IP address (Nicholas et al. 1999b).

The problem of many users sharing the address of a proxy server is apparent in the log records of the University of Illinois Library Web server. The client "www-2.library.uiuc.edu" appears often in the server logs, often requesting the same file many times in a single day. This client is the library's proxy server; any user with a cable modem or a non-UIUC dial-up account must log into the proxy server in order to access the Chemistry Library electronic reserves. At any given time, there may be several users accessing the electronic reserves through the proxy server, and the entries retrieved from the log are unable to differentiate individual user sessions.

Users with multiple IP addresses and IP addresses with multiple users are not the only difficulties in identifying users and their usage of the Web site. Even if each user had an individual and static identifier that could be used as identification every time that user accessed the server, the problem of defining *a transaction* would remain. Would *a transaction* be a request for and receipt of a single file (or error message), or would *a transaction* be all of the entries made by that user in some period of time? (Mobasher et al. 1997).

Two other complications in retrieving meaningful statistics from Web server log data exist: handling pages that have multiple components and handling pages that have been cached outside the server. The problem of pages with multiple components has been noted repeatedly in transaction log analysis studies (Mobasher et al. 1997; Nicholas et al. 1999b; Stehle 2002). In HTTP 1.0, when a client requests a URL from a server, the server opens a separate connection to deliver each component on the page. If the page is an HTML text page with one GIF-format image, two connections are opened, and two requests are noted in the server log: one for the HTML file and one for the GIF file. HTTP 1.1, the newer version of the protocol, is supposed to use one connection to deliver all the elements of a file, but this has not been the case with the UIUC Library Web server. Each of the class index pages in the Chemistry Library electronic reserves includes a GIF icon as a link to download Adobe Acrobat. Each time these pages are accessed, two requests appear in the logs: one for the index page and one for the icon. These accesses artificially inflate the usage statistics for the index pages.

At the other extreme of usage altering is the artificial deflation of usage counts caused by caching of files outside the Web server. This problem has also been noted by many researchers (Carter 1995; Fieber 1999; Goldberg 1995; Nicholas et al 1999b). When a document is stored in a local browser cache, it allows the user of that browser to view the document repeatedly without requesting it from the server each time. This eliminates some unnecessary network traffic, but it also means that the transaction logs do not record every usage of the document (Carter 1995). If the browser is on a public access terminal, several users could view the document, leaving no record of their usage in the server logs. Documents can also be cached by proxy servers, cache servers, or by information service providers, allowing users at different terminals to view the cached document. These caches "remove pieces of what was a sketchy record of user behavior to begin with" (Fieber 1999).

## *PUTTING IT ALL TOGETHER: OPTIONS FOR LOG ANALYSIS*

Given the redundant log information about pages with multiple components and the incomplete log information about pages that have been cached, what can be done to generate more meaningful and reliable usage statistics? Some researchers emphasize that "there is no single measure of consumption and each measure has to be taken with a large dose of statistical salt" (Nicholas, Huntington, Lievesley & Withey 1999a). Other researchers insist that meaningful statistics cannot be generated at all, and that "it is not enough to say that the statistics should be taken with a grain of salt; they should be taken with a salt lick" (Goldberg 1995).

It is true that server access logs, in their raw form, are not accurate site usage records, but measures can be taken to improve the usefulness of the data. These measures include altering the log format, applying commercial analysis software, requiring all users to log into the secure server, preventing pages from caching, and using custom intermediate log pages. Additional steps that could be taken at the UIUC Chemistry Library would alter the way the server data query form is used or setting up specialized direct SQL searches.

The first obstacle the Chemistry Library faces is the format of the logs themselves. Data that are not recorded cannot be analyzed. However, as Bauer noted, data recorded in the logs "are the ultimate limiting factor in what log analysis software can do for you" (Bauer 2000). Many of the commercial programs available are designed to analyze files in the Common Log File format, not database tables. With files in this format, libraries can follow the steps to analyze data outlined by Johnson:

1. Obtain log files from the library system administrator, determine their format, and install them on a local workstation.
2. Select, download (if necessary), and install analysis software, such as Analog®, wwwstat, Microsoft Usage Analyst®, WebTracker®, or WebTrends®.
3. Run the analysis software, generating the appropriate reports (Johnson 2000).

It is possible to configure IIS 4.0 to log files in Common Log File format, and it is possible to log to a file instead of a database. This would, however, cause great upheaval in the Library system, and there are other ways to manipulate or augment the data that is already available.

One of the features of many log analysis programs is an attempt to identify individual sessions and track the usage patterns of individual users. The Chemistry Library requires off-campus electronic reserves users to log into the system through a proxy server. If all users logged into the proxy server, records could be kept of their individual paths through the site. This approach presents several difficulties. First, it would be an annoyance to users. As Stehle notes, "signons/Ids/passwords may help meet the needs of marketers and service developers, but they are cumbersome and annoying from the user's point of view" (Stehle 2002). Second, because users may not log out of the proxy server, the problem of determining the end of an individual session would remain. Third, keeping the records would require a separate logging table, increasing network traffic and the amount of space needed for storage, because the current system configuration records the address of the proxy server in the access logs, rather than any identifying information about the individual logged into the proxy server. Finally, and most significantly, tracking individual users by personal identification (such as their University ID numbers) raises ethical concerns. While identifying information may be taken from users who check out traditional print reserve materials, the individual circulation records are not preserved after the items are returned to the library. (A running tally of how many times materials have been checked out is kept for each day, but the borrower names are not recorded.) Preserving the individual records of Electronic Reserve usage in a log file would be a significant change from the traditional statistics.

One way to improve statistics without requiring major changes in the server configuration or user habits is the prevention of document caching. It is possible to configure HTML and ASP documents with META tags that cause the page to expire from remote caches immediately. This would force browsers to reload the document from the server each time it was requested by a user.

This approach, however, would increase traffic on the network, which could conceivably cause slow transactions and frustrate users (Goldberg 1995). The current configuration of the Web server logs at UIUC and the search page used to access the log data report transactions that resulted in the error code 304, a message indicating that the user's cached copy was identical to the copy on the server. The document is not transferred from the server in this case, but there is a record of the request being made.

Another way to improve the Web usage data is to make use of additional data gathering methods especially those that focus on the user's motivations rather than his actions. As Yu and Apps stated, "the logged data reflect only what the user does but tells little about why. Therefore, to examine the causal factors behind use and non-use and the potential determinants for patterns of use, a great deal of contextual information still needs to be collected through other channels" (Yu & Apps 2000). These methods, including "user surveys, focus groups, and other feedback mechanisms, can gather user opinions on site content, navigation, look-and-feel, as well as assess user satisfaction and the reasons that users visited the site or navigated as they did" (Haigh and Megarity 1998). Many libraries have used interviews and questionnaires to assess patron satisfaction with their services. During the last year, the Undergraduate Library at UIUC and the Chemistry Library both surveyed students about their usage of electronic reserves. These surveys, especially the open-ended questions that gave students an opportunity to express their feelings in their own words, provide the libraries with an opportunity to discover what the students enjoy about electronic reserves as well as what needs to be improved (Chrzastowski 2002). These surveys provide the accountability and opportunity for service improvement that Johnson notes as reasons for performing usage analyses.

In addition to these qualitative evaluations, additional quantitative information can be generated through tracking programs that run on outside servers. Many of these services, such as NedStat® or SiteMeter®, can be used for free on individual personal homepages and there are similar devices for corporate sites. These programs are usually written in JavaScript and record information about users' paths through the site, types of Web browsers, operating systems, the referring page that led them to the site, and more. The hosting sites can generate easy-to-read tables of the information. A disadvantage of this approach is that the JavaScript can sometimes frustrate users by causing slow document loading and there are privacy concerns inherent in allowing an outside entity to log information about a library Web site and its users.

Another approach would be to record Web clicks on the same local server using custom script software. In this approach transaction information would be written to a file before redirecting the browser to the selected Web address. Several articles in this volume report applications of this technique.

Cookies (small data packets stored on users' computers when they access a Web site) can also be added to documents and used to identify individual users. This raises some of the same privacy concerns of requiring users to log into a server for access. It also assumes that all users will accept cookies and that the cookies will be kept on their hard drive, rather than discarding them immediately or before their next access.

For the Chemistry Library's immediate needs, there are two simpler methods of generating more meaningful statistics. First, it is possible to perform more sophisticated queries of the server log database through direct SQL queries. In our situation, and because queries must be performed by Library Systems staff, all queries must be prepared in advance and should not be requested very often.

For more frequent data collection, the online query form, accessible remotely by authorized staff, can also be manipulated to perform a variety of searches and the results of these searches may be combined. In order to eliminate the problem of multiple usage counts for a single page due to image elements, a search limited to HTML files can be performed. Alternatively, a search for all files can be performed, followed by a search limited to image files, and the second count can be subtracted from the first. Searches can also be limited to particular error codes: for example, a search for all requests that resulted in the code 304 would show how many accesses came from browsers that had already cached a copy of the requested document. Searches can be limited to particular IP addresses; this type of search could show how many accesses came through the proxy server relative to the total number of accesses.

The best solution may be a combination of options. Combining quantitative methods of server log analysis (in its various forms) with qualitative methods of analysis, such as user interviews or surveys, provides a more complete picture of Web resource usage. For example, the server log data was able to show that requests were made from off-campus users that resulted in a "Forbidden" error message. The Chemistry Library's electronic reserves survey gave students who did not use the electronic reserves (or who used them, but would have preferred not to use them) an opportunity to explain their reasons. Many surveyed students were able to explain that they had tried to access the materials directly, not knowing they were required to access the proxy server through a different page first. The contents of the log alone do not differentiate between students attempting to access the wrong location, indicating that access instructions should be clarified, and unauthorized users successfully blocked from accessing restricted materials. The log records how many authorized users were able to access the page, but it cannot record authorized users that it does not recognize.

## IN THE END, THERE ARE NO PERFECT NUMBERS

Web server logs appear, at first glance, to be vast mines of information about users. It seems that every action a user performs could be recorded and analyzed, and the findings used to improve the available resources. Of course, Web server logs only record usages of pages that are stored on that particular local Web server. Clicking on a link to a page on a different server does not generate any log entry on the local server. An analysis of complete transaction log data would generate an unmanageable amount of information of questionable value and raise significant concerns about user privacy. The recorded data are primarily intended for use in maintenance of the server, not for analysis of individual document usage. Nicholas et al. has stated that, "logs enable us to follow the progress of packs of users rather than individuals and to read the broad outlines of their information fingerprints. It is a large but fuzzy picture" (Nicholas, Huntington, Lievesley & Withey 1999a).

This picture can be clarified somewhat through manipulation of the data recorded and the addition of data from outside sources, but it will never be a perfect record of usage. To be meaningful, though, the record does not need to be perfect. Bauer notes that, "few measures of usage are [perfect]. For example, when we count people who come through the doors of our library, we don't know if they are there to read books or magazines, or just use the bathroom. When we circulate a book, we don't know why it was selected or even if it is read" (Bauer 2000). Libraries have kept usage statistics through a variety of methods for a long time; the fact that statistics tend to be approximations is already well known (Chrzastowski & Olesko 1997). As libraries increase their Web presence and offer more resources online, the very process of uncovering usage statistics and the attempt to apply familiar analysis methods in a new environment is important. Web transaction log research, in some cases, has "turned out to be the type of research where the journey itself proved to be more important than the information produced as a result of having cracked the code. This was because, by the very act of cracking the code, you are questioning the web itself" (Nicholas, Huntington, Lievesley, & Withey 1999a). While attempting to "crack the code" of transaction logs, libraries can discover new ways of looking at old practices, including usage analysis, that will lead to improved service for all users.

## REFERENCES

Bauer, K. 2000. Who goes there? Measuring library Web site usage. *Online 24(1).* from http://www.onlinemag.net/OL2000/bauer1.html. Accessed May 8, 2002.

Carter, D. S. 1995. Web server transaction logs project report. Available at http://www.personal.umich.edu/~superman/AP/Report.html. Accessed May 8, 2002.

Cochrane, P. A., & Markey, K. 1983. Catalog use studies–Since the introduction of online interactive catalogs: Impact on design for subject access. *Library & Information Science Research* 5:337-363.

Chrzastowski, T. E. 2001. Electronic reserves in the science library: Tips, techniques, and user perceptions. *Science & Technology Libraries* 20(2/3):107-119.

Chrzastowski, T. E., & Olesko, B. M. 1997. Chemistry journal use and cost: Results of a longitudinal study. *Library Resources and Technical Services* 41(2):101-111.

Fieber, J. 1999. Browser caching and Web log analysis. *ASIS Midyear Conference* May 25 Pasadena, CA. Available at http://ella.slis.indiana.edu/~jfieber/papers/bcwla/bcwla.html. Accessed May 8, 2002.

Flaherty, P. 1993. Transaction logging systems: A descriptive summary. *Library Hi Tech*, 11(2):67-78.

Goldberg, J. 1995. Why Web usage statistics are (worse than) meaningless. Available at http://www.goldmark.org/netrants/webstats/. Accessed May 8, 2002.

Haigh, S. & Megarity, J. 1998. Measuring Web site usage: Log file analysis. *Network Notes* 57. Available at http://www.nlc-bnc.ca/9/1/p1-256-e.html. Accessed May 8, 2002.

IBM. 2002. Log file formats. Available at http://support.accrue.com/hitlist/techdb/tdb-logs.htm. Accessed May 8, 2002.

Johnson, R. 2000. Librarian's guide to measuring web usage. Available at http://www.slis.ualberta.ca/cap00/rjohnson/. Accessed 8 May 2002.

Kaske, N. K. 1993. Research methodologies and transaction log analysis: Issues, questions, and a proposed model. *Library Hi Tech* 11(2):79-86.

Kurth, M. 1993. The limits and limitations of transaction log analysis. *Library Hi Tech* 11(2):98-104.

Lewenberg, A. 2002. Conversation with author. Urbana, Ill., May 8.

Mobasher, B., Jain, N., Han, E., & Srivastava, J. 1997. *Web mining: Pattern discovery from World Wide Web transactions* (Tech. Rep. No. 96-050). Minneapolis: University of Minnesota, Department of Computer Science.

Nicholas, D., Huntington, P., Lievesley, N., & Withey, R. 1999a. Cracking the code: Web log analysis. *Online & CD-ROM Review* 23(5):263-269.

Nicholas, D., Huntington, P., Williams, P., Lievesley, N., Dobrowolski, T., & Withey, R. 1999b. Developing and testing methods to determine the use of Web sites: Case study newspapers. *ASLIB Proceedings* 51(5):144-154.

Peters, T. A. 1993a. The history and development of transaction log analysis. *Library Hi Tech* 11(2):41-66.

Peters, T. A., Kurth, M., Flaherty, P., Sandore, B., & Kaske, N.K. 1993b. An introduction to the special section on transaction log analysis. *Library Hi Tech* 11(2):38-40.

Simpson, C. W. 1989. OPAC transaction log analysis: The first decade. In J. A. Hewitt (Ed.), *Advances in Library Networking* 3 Greenwich, CT: JAI Press.

Stehle, T. Getting real about usage statistics. Available at http://216.167.68.146/marketscope/conaghan/stehle.html. Accessed 8 May 2002.

Yu, L., & Apps, A. 2000. Studying e-journal user behavior using log files: The experience of SuperJournal. *Library & Information Science Research* 22(3):311-338.

# Challenges and Changes:
# A Review of Issues
# Surrounding the Digital Migration
# of Government Information

Robert Slater

**SUMMARY.** This paper presents a review of recently proposed policy changes for the dissemination of government information and a critical look at the arguments being made for and against an immediate migration to electronic-only dissemination and preservation of government publications. Issues of authenticity, security, permanence of records, access, format, and cost are examined. *[Article copies available for a fee from The Haworth Document Delivery Service: 1-800-HAWORTH. E-mail address: <docdelivery@haworthpress.com> Website: <http://www.HaworthPress.com> © 2001 by The Haworth Press, Inc. All rights reserved.]*

**KEYWORDS.** Government information, information dissemination, NCLIS, GPO

## INTRODUCTION

In the past two years, both the Government Accounting Office (GAO) and the National Commission on Libraries and Information Sci-

Robert Slater is Digital Information Services Librarian, Kresge Library, Oakland University, Rochester, MI (E-mail: rslater@oakland.edu).

[Haworth co-indexing entry note]: "Challenges and Changes: A Review of Issues Surrounding the Digital Migration of Government Information." Slater, Robert. Co-published simultaneously in *Science & Technology Libraries* (The Haworth Information Press, an imprint of The Haworth Press, Inc.) Vol. 21, No. 1/2, 2001, pp. 153-162; and: *Information Practice in Science and Technology: Evolving Challenges and New Directions* (ed: Mary C. Schlembach) The Haworth Information Press, an imprint of The Haworth Press, Inc., 2001, pp. 153-162. Single or multiple copies of this article are available for a fee from The Haworth Document Delivery Service [1-800-HAWORTH, 9:00 a.m. - 5:00 p.m. (EST). E-mail address: docdelivery@haworthpress.com].

http://www.haworthpress.com/store/product.asp?sku=J122
© 2001 by The Haworth Press, Inc. All rights reserved.
10.1300/J122v21n01_13

ence (NCLIS) have published reports that encourage an all-digital dissemination of government information (GAO 2001; NCLIS 2002). These reports have created a great deal of controversy. Critics question the wisdom of having only electronic dissemination of information, and have raised many valid concerns. Of the two reports, the NCLIS proposal has generated the most discussion. It is the more sweeping and in-depth, going so far as to suggest regulatory changes to ensure a comprehensive move to digital only dissemination. The proposal points out that the very existence of widely popular digital channels for information dissemination (the Internet and Web, among others) has already produced a situation where a large number of government publications exist in a dual print and electronic existence and many exist solely in digital format. More importantly, the NCLIS report points out that many of the less well-known and used publications which are published solely on the Web, only to disappear forever a few short months after their inception. Because the current polices and regulations regarding the distribution of government information have done little to address these types of items, many are created and cease in digital format only. The need to preserve these valuable sources of information necessitates a vast overhaul in the way government agencies disseminate and archive information. The NCLIS proposal is forward looking–to an age of all digital dissemination; but it also contributes to the current problem–the age of digital dissemination of government information is upon us and unless the digital big picture is addressed now, a vast amount of information may be lost forever. The NCLIS proposal and the literature and discussion it has generated, has brought into clear focus six major issues that must be addressed in the government move to a digital future: authenticity, security, permanence of records, access, format, and cost. This paper focuses on these challenges and the problems they present, and discusses possible solutions.

## *NCLIS*

The most recent and sweeping proposals to the dissemination of government information were presented by the National Commission on Libraries and Information Science (NCLIS) report, "A Comprehensive Assessment of Public Information Dissemination: Final Report." Little change may be brought about by this proposal, given the Bush administration's failure to continue funding the commission (Flagg 2002). The assessment the commission made is a comprehensive look at the challenges raised by the current and future condition of the United States government's digital dissemination of information. It is an unfortunate truth that the rates at which electronic information technologies have been evolving give even the most technologically-oriented information specialists pause from time to time. The intricacies of networked information systems are complex enough that entire teams of specialists in various fields must work closely together to produce information systems that are reliable and useful. The concept here is that a team is required to handle complex problems related to information technology. A team consists of members with complementary skills necessary for the successful completion of IT projects. These teams rely on individual members to manage and explain emergent technologies related to the project they are working on (Peled 2000). The uncertainties of the digital future strongly contribute to government documents librarians' dissatisfaction with the idea of relinquishing their physical collections and relying on electronic access.

There is real concern that information will be lost as the equipment and software used to create and use digital information becomes obsolete. Depository librarians and others feel that in a rapidly changing electronic environment, they cannot be sure that the currently available electronic files will be able to be opened and used in the future or that there will be an easy, cost-effective means to migrate this generation of electronic products to future formats and media (NCLIS 2002).

The traditional collocation of information has provided a location for accessing it. How will libraries offer traditional services of selecting, procuring, organizing, and providing access to items when there is no physical medium? As the Government Documents Roundtable (GODORT) of the American Libraries Association (ALA) stated, "NCLIS should re-emphasize the role of depository librarians and libraries in its report" (Miller 2001). Archivists and preservationists make the argument that paper is permanent and electronic files subject to technological obsolescence and complex access issues and the concomitant notion that "we must guard against letting the digitized version totally supplant the original" (Ojala 2001). As with other digitized collections, there are archival concerns that many documents may not be upgraded as hardware and software evolve. The digital future is inevitable and the preparations for meeting it would best be addressed by librarians. To ensure the creation of a system that libraries will find useful, librarians must be among the information professionals who do more than point out the problems of digital systems–they must also propose solutions to these problems.

## KEY ISSUES OF DIGITALIZATION

There are several key points that must be addressed and these have already been raised in the government documents and library literature. The two most discussed concerns are digital dissemination and the proposal of merging the National Technical Information Service (NTIS) and the Federal Depository Library Program (FDLP) into several new agencies headed by the Public Information Resources Administration (PIRA) managed under the authority of the Library of Congress (LC). Issues that are being addressed include the following.

### Authenticity

How does one verify that an electronic record they are receiving or viewing is the "official" version of a document? Electronic documents, when compared to their print counterparts, can be changed or modified, accidentally or deliberately, with little indication to an end user, other than a user's prior knowledge that changes have been made. Government documents have unique cases of intentional changes to the text of records. Consider the frequent text changes in *The Congressional Record* by legislators to revise the meaning of what s/he said on the floor (Peterson 2001). Once the *Congressional Record* is printed and distributed, no easy method for changing the historical evidence is available to anyone. It would require tracking down and editing or destroying thousands of copies of the entry in question. In a digital format, particularly on a centrally located data server, it would be possible, although extremely difficult, for someone to permanently alter the historical record. Digital watermarking technologies of varying strengths allow files to be marked in such a way that a quick scan for the embedded digital watermark would show the file had been altered (Acken 1998). When a file has been watermarked in this way it is invisible to the user, but not to a program designed to check the file before distributing it or serving it to a user (Yeung 1998). Having many distributed copies, including off-line archival back-ups, when coupled with a watermarking technology, would allow the discovery of any alteration of files. Any altered files could be then deleted and replaced with a version of the file that has been verified to be unaltered.

### Security

Related to the issue of authenticity is the issue of security. Potentially more harmful than an interested and knowledgeable party's ability to alter records for their own benefit (or another's harm), are malicious attempts at destroying digital records in bulk. Anyone who has used the existing bulk of electronically disseminated government information is familiar with computer viruses attacking government servers and damaging data located on them. In the current FDLP model, the distributed architecture insures depositories which could handle all but the most horrific disaster.

## Permanence of Records

Closely tied to the issue of security is that of permanence and the capability to use archived materials in an electronic environment. The physical media that stores digital information changes rapidly. For example, within the past 20 years digital formats have changed from 9-track magnetic tapes to 5.25" floppy diskettes to the current DVD technologies. In general, rapidly changing formats and outdated retrieval methods do not have an equivalent situation in a paper-based collection, as emphasized by archivists criticizing the government's move in a digital direction. Permanence of records and the ability to migrate entire collections to new storage formats without compromising or losing data should be guaranteed before migrating to an all-digital collection.

## Access

Accessing electronic records requires computer hardware, software and, in the NCLIS proposed central depository design, Internet or network access. According to a recent United States Census survey, 56.5% of American households have a computer. Slightly fewer households, 50.5%, have Internet access, so 6% of American households have computers, but have not yet been connected to the Internet (NCLIS 2002). Remote electronic depositories create additional access complications for users and information professionals alike. Resolving access problems can be very time consuming and difficult to understand by many citizens. Rationale may be that more people would be able to access the information freely from their homes, but this can also create additional technological issues that need to be addressed.

## Format

Related to permanence is the decision of which format electronic data will be stored. If all multi-media formats (film, interactive computer programs) and non-print formats (maps, images, sound recordings) were excluded and focus was on text documents, there is still the decision of format type in which to save and present data. Current format types include Adobe Portable Document Format (PDF), ASCII text, 16 or 32 bit UNICODE, HTML (version 1-4), or XML DTD with attached CSS or XSL-FO. The myriad formats makes this decision non-trivial, even when focusing on print formats. These format decisions preclude non-print media formats.

## Costs

Each of the aforementioned issues has a direct effect on costs. It may be more cost effective for government agencies to employ a centrally stored collection of electronic records, thus reducing the distribution costs of print records, but does it provide better access for citizens? It is difficult to ascertain whether or not the FDLIP and NTIS distribution systems that are already in place would be more cost effective than an electronic-only system. The proposal by the GAO to migrate to an electronic dissemination system does not specifically address cost effectiveness. Michael DiMario, the Public Printer at the time both the GAO and NCLIS studies were released, faulted the GAO report for not providing the comprehensive study of the impact of providing documents solely in electronic format it claimed to because it only briefly mentions cost considerations (DiMario 2001). The Commission went on to defend its brief consideration of costs by noting that their purpose was to focus their study on the impact of providing government information solely in electronic format and on various issues concerning the feasibility of transferring the depository library program to the Library (GAO 2001).

The Government Accounting Office is unable to give definitive evidence that an electronic system would be more cost effective than the current one, yet, interestingly, the document opens with the statement "Electronic dissemination of government documents offers the opportunity to reduce the costs of dissemination and make government information more usable and accessible" (GAO 2001). A report from the Government Printing Office states "while there are many benefits inherent in the use of electronic information, including more timely and broader public access, there is no conclusive data at this time to support the assertion that it will result in significant savings to the program as a whole in the next few years" (GPO 2002).

## Putting It All Together

Authenticity, security, and permanence issues are closely interrelated and are dependent upon the founding concepts of how to electronically store government information. A system similar to OCLC's geographically distributed system of servers and redundant back-ups would go a long way to addressing some of the aforementioned concerns. This system could be achieved with a clone system of the main "parent" government server located at each of the state depository libraries. The system could employ a hybridized form of the information harvesting approach to allow the parent server to upload copies of new data or documents found on regional servers and allow the regional servers to query the parent server for any new publications needed at their site. This system would result in at least 51 geographically distributed copies of the data and allow each regional depository library to contribute to and help maintain the historical archive. This structure also addresses security issues by providing 50 different servers.

Procedures for safeguarding against incursions into servers and the manipulation of the data on them have already been firmly established by banks and government agencies such as the Internal Revenue Service (IRS) and the Federal Bureau of Investigation (FBI). Even if we can insure that the data is authentic and redundant across many servers, it is an inevitable fact that the most carefully developed system will need a complete data-format conversion periodically. With the current increases in the size and speed of memory and central processing units by the time such a conversion was needed it would take a dedicated server farm a relatively short amount of time to accomplish automatic format conversions from the outdated formats to the more current ones. The need to decide on what specific file formats to use and the accompanying need for eventual format translation are challenging issues. These issues, however, can be dealt with by requiring that each government publication should be accompanied by a least-common-denominator version. This could be a simple text file format that describes, as much as possible, the resource that it is to accompany. Having a least-common-denominator version addresses not only the problems of access, but permanence as well. Consider that many electronic publications are provided in the proprietary Adobe PDF format, which requires special software to view it. For a text presentation, the least-common-denominator text version would be a non-marked up version of the text (no HTML, XML, or similar encoding), accompanied by a standardized metadata description of the resource. In the case of multi-media files, the least-common-denominator version could be a simple metadata file, created using current or new metadata encoding techniques (such as Dublin Core), which can be arbitrarily rich. For audio-visual and multimedia materials, this would require that the text of any words spoken or displayed in the item, along with brief descriptions of the visual/non-speech elements, be included in a text file that accompanied the resource. Thus, even if it should become feasibly impossible in the future to view motion picture graphics (MPEGs), the least-common-denominator file accompanying that MPEG would still preserve access to the content of those MPEG. The least-common-denominator formats for various electronic media could be re-evaluated and updated as often as needed. This approach allows government information producers to experiment with new technologies that speak directly to the needs of the citizens who use the resources while still ensuring some stability to the historic record.

## CONCLUSION

A great deal of concern has been expressed over the conversion of government information to an all-electronic form. Michael DiMario stated "the day is coming when Federal Government information may be made available to the public solely in electronic format . . . that day is not here yet nor is it likely to appear in the foreseeable future" (DiMario 2001). Many information professionals have expressed similar concerns while also acknowledging the benefits of electronic access. When technological changes prompt policy changes as those proposed by the NCLIS, librarian's responsibilities are to insure that appropriate safeguards are in place and that the critical issues of archiving and access are being addressed. Librarians, as information specialists, must carefully evaluate these ongoing issues and propose the best possible solutions.

The NCLIS proposal does not abolish depository libraries across the country. Under the newly proposed agencies, FDLP libraries would become PIRA libraries, filling a role that would be essentially the same as before, except now they would direct patrons to electronic resources that meet their information needs–something that many depository librarians are already doing. Under the proposed new system, a centralized access point and archive would be available for the electronic information resources. The vast historical record that already exists in government depository libraries must also be considered. Most of these historic print records will never be converted into a digital format, ensuring that there will always be a place for the historic print archive of government information in libraries.

Providing information in electronic format inevitably raises concerns about patron access to this information. Getting access to government information has always required effort on the part of the individual by going to a depository library and finding, or getting assistance to find materials. Having the information available electronically, instead of in print, provides its own unique barriers to access. In the next 10 years the proposed system will cost at least as much, if not more, to implement than simply perpetuating the current system. However, the foundation for an electronic infrastructure for the government and public is currently being laid. A reliable system of electronically distributed government information will most likely be more cost effective than the current print distribution system, even factoring in the cost of off-line archiving, periodic data format conversion, and equipment purchase and maintenance.

A migration to electronic dissemination of government information will only increase access to government information, not decrease it. More and more, patrons of libraries are already demanding electronic access to items that were traditionally provided in print, even to the point of ignoring resources that might be appropriate to their information needs in favor of less useful ones that are available at the click of a mouse. In a recent three year study conducted by the Ohio State University, it was found that "while e-journal usage was increasing, there was an accompanying decrease in the use of printed journals" (Rogers 2001). As information seekers become more technologically proficient and dependent, they will come to appreciate the efforts that are being made now to provide permanent and efficient access to government information.

## REFERENCES

Acken, John M. 1998. How watermarking adds value to digital content. *Communications of the ACM* 41(7): 74-77.

Barnum, George. 2002. Availability, access, authenticity, and persistence: Creating the environment for permanent public access to electronic government information. *Government Information Quarterly* 19(1): 37-43.

Cox, Wendell and Love, Jean. 1996. 40 Years of the US Interstate Highway System: An Analysis The Best Investment A Nation Ever Made. Available at http://www.publicpurpose.com/freeway1.htm, accessed June 5, 2002.

Department of Transportation and Related Agencies Appropriations Act, 2000 Public Law 106-69, 106th Congress.

DiMario, Michael F. 2001. Comments from the Government Printing Office. In Appendix IX of Information Management: Electronic Dissemination of Government Publications. March 30. GAO-01-428 http://www.gao.gov/new.items/d01428.pdf, accessed May 28, 2002.

Flagg, Gordon. 2002. Bush Budget Boosts Librarian Recruitment, Kills NCLIS. *American Libraries* 33(3).

Government Accounting Office (GAO). 2001. Information Management: Electronic Dissemination of Government Publications. March 30. GAO-01-428 http://www.gao.gov/new.items/d01428.pdf, accessed May 28, 2002.

Miller, Ann. From the Chair. 2001. http://sunsite.berkeley.edu/GODORT/columns/chr_200103.html, accessed May 30, 2002.

National Telecommunications and Information Administration, U.S. Department of Commerce (NCLIS). 2002. *A Nation Online: How Americans Are Expanding Their Use of the Internet.* Available at http://www.ntia.doc.gov/ntiahome/dn, accessed May 30, 2002.

Ojala, Marydee. 2001. Preservation, Conservation, and Copyright Infringement. *Online* 25(5).

Order, Norman. 2001. GAO: Digital Docs Raise Questions. *Library Journal* 126(8): 15-16.

Peled, Alon. 2000. Creating winning information technology project teams in the public sector. *Team Performance Management* 6(1/2): 6-14.

Perritt, Jr, Henry. 2001. NCLIS assessment of public information dissemination: Some sound ideas tarnished by defense of obsolete approaches. *Government Information Quarterly* 18(2): 137-140.

Peterson, Karrie, Cowell, Elizabeth, and Jacobs, Jim. 2001. Government Document at the Crossroads. *American Libraries* 32(8): 52-55.

Rogers, Sally A. 2001. Electronic Journal Usage at Ohio State University. *College & Research Libraries* 62(1): 25-34.

Shuler, John. 2002. Libraries and government information: The past is not necessarily prologue. *Government Information Quarterly* 19(1): 1-7.

U.S. Government Printing Office. 1996. Report to the Congress: Study to Identify Measures Necessary for a Successful Transition to a More Electronic Federal Depository Library Program. *GPO Publication 500.11.* Available at www.access.gpo.gov/su_docs/dpos/rep_cong/wp/summary.wp, accessed May 29, 2002.

U.S. National Commission of Libraries and Information Science. A Comprehensive Assessment of Public Information Dissemination: Final Report Volume 1 Washington, DC:US. http://www.nclis.gov/govt/assess/assess.html.

Yeung, Minerva M. 1998. Digital watermarking. *Communications of the ACM* 41(7): 30-33.

# Index

Academic libraries. *See also* Geology
    librarianship;
    Science/technology libraries
  changing role of, 88-89
  decreasing on-site use of, 4-5
  electronic reserves and, 5
  historical funding of, 6
  meteorologists' suggestions for, 60-62
  science/technology libraries and, 4-5
  unsustainability of, 5-6
Access, electronic resources and, 157
Access in perpetuity, 22
"Appropriate copy" problem, 110-111
Archiving, 22-23,24
  electronic journals, 71-72
  electronic resources, 76
  for Engineering Documents
    Collection, UIUC, 40-41
*The Arctic Bibliography,* 77
Article citation issues, for electronic
    journals, 72
Articles, challenges of awareness of, 74
Association of Research Libraries
    (ARL), 121
Astrophysics Data System (ADS), 72
Atmospheric scientists. *See*
    Meteorologists
Authenticity, electronic resources and,
    156
Awareness, challenges of, 74

Bibliographic classification systems,
    as obstacle to intellectual
    access, 49-50
Biomechanics, 49

BioOne, 22
Books. *See* Electronic books
Burright, Marian, 21
Bush, Vannevar, 66-67

Cataloging of electronic resources, 75-76
Center for Research Libraries (CRL),
    24-25
Centralization. *See* Mergers
Chemistry Library, UIUC
  generating meaningful and reliable
    Web usage statistics for,
    146-150
  missing information and, 144-146
  monitory use of Web site at, 141-142
  quantifying usage at, 140
  transaction log analysis at, 142-144
Citation issues, for electronic journals,
    72
Clarkson University, 47
Classification systems, as obstacle to
    intellectual access, 49-50
Clearinghouses, 30
CNRI Handle System, for DOI, 102-104
Collection development,
    cross-disciplinary areas and,
    50-51
*Collective Bibliography of North
    Dakota Geology,* 77,81
Common Log Format, HTTP, 143-144
CONTENTdm Multimedia Archival
    Software, 33-35
Copyright law, 79-80
Core collecting, 24
Cost issues, for electronic resources, 158

Cross-disciplinary, defined, 47-48
Cross-disciplinary areas, 50-51
  optimal library organization for, 51
CrossRef, 22,108-110
Custom script software, 149


Data sets, locating actual, 74-75
Departmental libraries
  characteristics of, 6-7
  merging, 6-8
Dewey Decimal classification system, 49
Digitalization
  access issues and, 157
  authenticity issues for, 156
  cost issues for, 158-159
  format issues for, 157-158,159-160
  geology librarianship trends for, 68-74
  government information issues of, 156-160
  permanence of records issues for, 157
  science/technology library trends for, 7
  security issues for, 156-157
Digital Library Federation, 23
The Digital Library for Earth System Education, 77
Digital Object Identifiers (DOIs), 22,72
  advantages of, 99-101
  assignment of, 112-113
  background information for, 98-101
  defined, 99
  displaying, 113-114
  Handle System for, 102-104
  metadata for, 104-106
  number structure for, 101-102
  policies for, 106-107
  reference linking and, 114-116
  resolution for, 102-104
  sci-tech publishers and, 111-116
DiMario, Michael, 158

DOIs. *See* Digital Object Identifiers (DOIs)
DOI-X, 108
Drake, Miriam, 4
Dspace, 71
Dublin Core, 75,160


E-journals. *See* Electronic journals
Electronic books, 73
Electronic collections. *See also*
    Government information
  acquisition process for, 19-21
  archiving, 24
  delivery considerations for, 26-27
  free public usage and, 20
  licensing issues and, 20-21
  need for simple interfacing of, 26
  quality control and, 20
  traditional library concerns for, 21-23
  at University of Notre Dame, 17-19
Electronic journals, 68-73
  accessing, 69-70
  advantages of, 68-69
  archiving, 71-72
  article citation issues and, 72
  assignment of DOIs to, 112-113
  challenges of awareness of articles in, 74
  cost challenges of, 70-71
  disadvantages of, 69
  format issues for, 72-73
  indexing, 69,70
  portals for, 70
  refereed, 71
  reference linking of, 107-110
  standardization of, 70
  use data for, at Grainger Engineering Library, 136
Electronic reserves
  academic libraries and, 5
  collecting Web usage statistics for, 140-141
Electronic resources. *See also*
    Government information

access issues for, 157
authenticity issues for, 156
cataloging, 75-76
cost issues for, 158
determining use of, with library
    public terminals, 128-129
format issues for, 157-158
need for determining use of, 120
obstacles to deterring use data for,
    120-121
permanence of records issues for,
    157
preservation of, 76
Purdue University project for
    collecting use data for,
    121-125
security issues for, 156-157
Elsevier, 23
E-Metrics study, by Association of
    Research Libraries, 121
Engineering Documents Collection,
    UIUC. *See also* Grainger
    Engineering Library, UIUC
archiving considerations for, 40-41
digital portal to, 31-32
scanning workflow for, 36
software considerations for scanned
    images for, 36-40
sources of digital images for, 32
Web interfaces for, 33-36
Engineering education, history of, 44-47
Engineering libraries, 44
Engineering Library, University of
    Louisville, 10-11
Environmental engineering, 48


Federal Library Depository Program
    (FLDP), 79
Format issues, for electronic materials,
    157-158,159-160
Fracture, defined, 48
Free public usage, electronic
    collections and, 20
Fretwell-Downing Company, 70

Geographical Information Systems
    (GIS), 73-74
Geology librarianship. *See also*
    Academic libraries;
    Science/technology libraries
archiving issues for, 76-77
challenges of current awareness and,
    74
copyright and legislation issues and,
    79-80
digitalization trends in, 68-74
    electronic books, 73
    electronic journals, 68-73
    Geographical Information
        Systems, 73-74
digital projects and, 80-81
driving forces in, 67-68
finding data sets and, 74-76
government agencies and, 78-79
information literacy for, 78
reference tools for, 77-78
Georgia Institute of Technology, 47
*Global Change Master Directory*
    (NASA), 74-75
Government Accounting Office (GAO),
    153-154
Government information. *See also*
    Electronic resources
digitalization issues for, 156-160
Government Accounting Office
    and, 153-154
National Commission on Libraries
    and Information Science
    proposals for, 153-155
Grainger Engineering Library, UIUC.
    *See also* Engineering
    Documents Collection, UIUC;
    Use data
recording user resource selection at,
    129
    methodology for, 129-132
    resource usage at, 134,244
    analysis of, 133-136
use of electronic journals at, 136
Grey literature, 30
Grounded theory, 55

Haank, Derk, 23
Handle System, for DOI, 102-104
Hawkins, Brian, 5
Hunter, Karen, 23
HyperText Transfer Protocol (HTTP),
    collecting Web usage
    statistics and, 143

IDF. *See* International DOI Foundation
    (IDF)
Indexing, electronic journals, 69,70
Information
    achieving intellectual access to,
        47-49
    achieving physical access to, 50-51
    obstacles to accessing, 49-59
Information literacy, 26
    for geology libraries, 78
Information-seeking behavior, 55
Intellectual access, to information
    achieving, 47-49
    obstacles to, 49-50
International Coalition for Library
    Consortia (ICOLC), 121
International DOI Foundation (IDF),
    99-101,107

John Crerar Library, The University of
    Chicago, 10
Journals, awareness of published
    material in, 74. *See also*
    Electronic journals
Just-in-time delivery, 25

*KGS Online Bibliography of Geology*
    (Kansas Geological Survey), 77

Libraries. *See also* Academic libraries;
    Science/technology libraries
    core collecting and, 24
    joint projects with publishers and, 23
    new challenges for, 25-27
    publicity and, 26

Library events. *See* Science Media Fair
    (S&E Library, UCSC)
Library instruction, 77
Library of Congress classification
    system, 49-50
Library terminals, obtaining use data
    from, 128-129. *See also*
    Grainger Engineering
    Library, UIUC; Use data
Licensing issues, electronic collections
    and, 20-21
Licklider, J. C. R., 66-67

Marketing. *See* Science Media Fair
    (S&E) Library, UCSC
Matylonek, John, 18-19
Mergers, 6-8,13-14
    approaches to, 9-10
    assessing viability of, 7-8
    benefits of, 9
    and closure of departmental
        libraries, 8-9
    decision making process for, 12-13
    examples of, 10-12
    factors determining, 12-13
    at Michigan State University, 11-12
    prevalence of, 11-12
Metadata, for DOI, 104-106
Meteorologists
    information gathering study of, 56-57
    information needs and problems of,
        57-60
    information-seeking behavior of, 55
    suggestions by, for academic
        librarians, 60-62
Michigan State University, library
    mergers at, 11–12
*MyGateway* portal, 70
*MyLibrary* portal, 70

National Commission on Libraries and
    Information Science (NCLIS),
    153-155

National Science Foundation, 44,45, 46-47
Natural Science Library (University of
    Louisville), 10-11
North Carolina State University, 70
North Dakota State University
    Libraries, 70,77


Oliver, Jim, 1
Online Public Access Catalog (OPAC),
    75
OpenURL, 111
Outreach efforts, library. *See* Science
    Media Fair (S&E Library,
    UCSC)
Overload, defined, 48-49


Permanence of records, electronic
    resources and, 157
Persistent Uniform Resource Locators
    (PURL), 76
Physical Review Online Archive
    (PROA), 72
Portals, for electronic journals, 70
Preservation. *See* Archiving
Promotion strategies. *See* Science
    Media Fair (S&E Library,
    UCSC)
Proximity, 7
Publicity, libraries and, 26
Public usage, free, and electronic
    collections, 20
Publishers
    DOI and, 111-116
    joint projects with libraries and, 23
Purdue University, project for
    collecting use data at, 121


Quality control, electronic collections
    and, 20
Queries, for server logs, 149

Refereed e-journals, 71
Reference linking, 107-110
    DOI and, 114-116
Reference tools, for geology libraries,
    77-78
Resolution, for DOI system, 102-104


Scholarly publications, changes in, and
    libraries, 5
Scholarly Publishing and Academic
    Resources Coalition
    (SPARC), 22,71
Science & Engineering (S&E) Library,
    UCSC, 88
    promotion strategies for, 93-94
    Science Media Fair at, 88-90
        promotion strategies for, 93-94
        10th Anniversary Seminar at, 90-93
Science Media Fair (S&E Library,
    UCSC), 88-90
    promotion strategies for, 93-94
    *vs.* 10th Anniversary Seminar, 91-93
Science/technology libraries. *See also*
    Academic libraries; Geology
    librarianship
    academic libraries and, 4-5
    centralization of, 6-8
    changes in technology and
        publishing and, 4
    digitization of, 7
    reasons for closing, 8-9
    suggestions for, 60-62
Scientific/technology publishers, DOI
    and, 111-116
Security, electronic resources and,
    156-157
Server logs, 143-144. *See also*
    Transaction log analysis
    missing information and, 144-146
    querying, 149
SPARC. *See* Scholarly Publishing and
    Academic Resources
    Coalition (SPARC)
Statistics. *See* Use data; Web usage
    statistics

Technology libraries. *See*
        Science/technology libraries
Terminals, library, for determining
        use of electronic resources,
        128-129. *See also* Grainger
        Engineering Library, UIUC;
        Use data
Terminology, as obstacle to
        intellectual access, 49
Tracking programs, 149
Transaction log analysis, 122-123,135.
        *See also* Use data; Web
        usage statistics
    history of, 142-143
    server logs records, 143-144


Uniform Resource Locators (URLs), 98
*Union List of Field Trip Guidebooks
        of North America Online*
        (Geoscience Information
        Society), 77
United States Geological Survey
        (USGS), 78-79
University of California at Santa Cruz
        (UCSC). *See* Science &
        Engineering (S&E) Library,
        UCSC
The University of Chicago, library
        mergers at, 10
University of Cincinnati, library mergers
        at, 11,12
University of Illinois at
        Urbana-Champaign (UIUC).
        *See* Chemistry Library,
        UIUC; Engineering
        Documents Collection,
        UIUC; Grainger Engineering
        Library, UIUC
University of Louisville, library
        mergers at, 10-11
University of Notre Dame
    electronic collections at, 17-19

just-in-time delivery at, 25
University of Washington Libraries, 70
Unpublished reports, 30. *See also*
        Engineering Documents
        Collection, UIUC
Use data, 120. *See also* Grainger
        Engineering Library, UIUC;
        Server logs;
    Transaction log analysis; Web
        usage statistics
    formulating standards for, 121
    from library public terminals,
        128-129
    obstacles to determining, 120-121
    Purdue University project for
        collecting, 121-125
    quantifying, 140
    vendors and, 120-121


Vendors, use data and, 120-121
Virginia Polytechnic Institute, 7


Wayne State University, 4,8
Web tools
    for accessing e-journals, 69-70
    for information literacy, 78
Web usage statistics. *See also* Server
        logs; Transaction log
        analysis; Use data
    for electronic reserves, 140-141
    generating meaningful and reliable,
        146-150
    HTTP and, 143
    perfect numbers for, 150-151
    quantifying, 40
    reasons for collecting, 141-142


Zar, Kathleen, 10
*Zportal,* 70

Printed and bound by CPI Group (UK) Ltd, Croydon, CR0 4YY

17/10/2024

01775687-0001